E-HEALTH TWO-SIDED MARKETS

E-HEALTH TWO-SIDED MARKETS

Implementation and Business Models

Edited by

VIVIAN VIMARLUND

Linköping University, Linköping, Sweden;
Jönköping University, Jönköping, Sweden

Amsterdam • Boston • Heidelberg
London • New York • Oxford • Paris • San Diego
San Francisco • Singapore • Sydney • Tokyo
Academic Press is an imprint of Elsevier

Academic Press is an imprint of Elsevier
125 London Wall, London EC2Y 5AS, United Kingdom
525 B Street, Suite 1800, San Diego, CA 92101-4495, United States
50 Hampshire Street, 5th Floor, Cambridge, MA 02139, United States
The Boulevard, Langford Lane, Kidlington, Oxford OX5 1GB, United Kingdom

Notices
Knowledge and best practice in this field are constantly changing. As new research and experience
broaden our understanding, changes in research methods, professional practices, or medical
treatment may become necessary.

Practitioners and researchers must always rely on their own experience and knowledge in evaluating
and using any information, methods, compounds, or experiments described herein. In using such
information or methods they should be mindful of their own safety and the safety of others,
including parties for whom they have a professional responsibility.

To the fullest extent of the law, neither the Publisher nor the authors, contributors, or editors, assume
any liability for any injury and/or damage to persons or property as a matter of products liability,
negligence or otherwise, or from any use or operation of any methods, products, instructions, or ideas
contained in the material herein.

British Library Cataloguing-in-Publication Data
A catalogue record for this book is available from the British Library

Library of Congress Cataloging-in-Publication Data
A catalog record for this book is available from the Library of Congress

ISBN: 978-0-12-805250-1

For Information on all Academic Press publications
visit our website at https://www.elsevier.com

Publisher: Mica Haley
Acquisition Editor: Rafael Teixeira
Editorial Project Manager: Mariana Kühl Leme
Production Project Manager: Chris Wortley
Designer: Maria Inês Cruz

Typeset by MPS Limited, Chennai, India

CONTENTS

Part I   Introduction to the Ecosystem for Two-Sided Markets, Barriers and Facilitators

Part II   Usability and Design

Part III   Safety and Privacy

Part IV   Implementation and Introduction of Technologies

Part V   Business Models

Part VI   The Future of the Area

LIST OF CONTRIBUTORS

P. Bertelsen
Aalborg University, Aalborg, Denmark

E.M. Borycki
University of Victoria, Victoria, BC, Canada

L. Botin
Aalborg University, Aalborg, Denmark

M. Eurich
ETH Zurich, Zurich, Switzerland; University of Cambridge, Cambridge, United Kingdom

A.W. Kushniruk
University of Victoria, Victoria, BC, Canada

C.E. Kuziemsky
University of Ottawa, Ottawa, ON, Canada

T. Mettler
University of Lausanne, Lausanne, Switzerland

J. Mitchell
OCAD University, Toronto, ON, Canada

C. Nøhr
Aalborg University, Aalborg, Denmark

P. Nykänen
University of Tampere, Tampere, Finland

P.S. Ruotsalainen
University of Tampere, Tampere, Finland

J. Treviranus
OCAD University, Toronto, ON, Canada

V. Vimarlund
Linköping University, Linköping, Sweden; Jönköping University, Jönköping, Sweden

FOREWORD

We have chosen for several reasons to publish a book on the area of e-health with special focus on issues related to two-sided markets and their effects. One is because this is the first edition of a book that takes an innovative approach toward explaining the reasons that the market for e-health needs to evolve and become a sustainable one, at least in Europe. Another reason is because of the need to sample knowledge and to find and design a market for e-health in which each side, customers and users, interacts with the other.

This is a book about two-sided e-health markets and the need for sustainability and renewal of the area in order to stimulate the market of e-health. It does not claim to encompass issues related to innovation policy, but what it does provide is a framework that explains what is needed to foster growth and renewal of the area.

As many books focus on evidence-based cases or on discussing the role of ICT in the area, this book gives more attention to fostering new growth alternatives and into identifying challenges that cut across industrial and service sectors and e-health. There are global challenges of international concern whose solutions must be sought specifically on the local level.

We expect this book to be of interest to healthcare managers, decision makers, stakeholders, researchers, patients, and citizens, and we hope to have the opportunity to receive comments and feedback from the research community and to be able to improve and extend the ideas suggested in the different chapters.

We further expect that future versions of this book and its chapters will also be published in electronic form, and that the next version of this book could indeed be written via an open innovation challenge, whereby we invite scientists, practitioners, and other domain experts to share their ideas with us. We hope that our current suggestions and ideas become a reality, and that these can contribute toward innovating the e-health area independently of whether they today are financed by taxes, subsidies, private initiatives, or other governmental alternatives.

PREFACE

For the last decades, financial, political, and demographical developments have changed institutional healthcare context worldwide. Overall, healthcare systems have been profoundly reformed from public owned healthcare systems to more market-oriented policy models (Helderman and Van der Grinten, 2005). Based on the idea of managed care or regulated competition, these two health systems strive after to stimulate competition between health providers and healthcare users (Enthoven and Van de Ven, 2007). In parallel, healthcare organizations are confronted with the introduction of new technologies and e-health services, new ways to interact with patients, practitioners, and stakeholders, new public regulations and the need to develop innovative business models and sustainable implementation models that allow the emergence of an ecosystem that stimulates collaboration, reduces redundant work, lowers costs, and increases market reach and penetration internationally. In doing so, new players will take their place in the ecosystem such that a rich milieu of options is supported and created, which is needed to bring individuals of different sides of a market together.

The importance of innovations in e-health context has become a key issue due to dynamics in customer demand, faster time-to-market, increased competition and the possibilities of co-creation in value networks using external ideas—often referred to as open innovation and crowd sourcing. Further, the possibility to innovate services either outside or inside the ecosystem of a two-sided market requires cross-domain engineering, open and linked data. On the one hand, the possibility is needed to make accessible internal data for sharing service ideas; on the other hand, one could use the external knowledge and innovation components from third parties and use this knowledge in the development of new services. However, in the health and care service sector, independent ecosystem members provide digital services, where they deliver additional value for both service consumers and other service providers. Service providers do not necessarily provide a complete service for consumers but can just render a specific part of a composite service.

It is interesting to note that, while the e-health area has given priority to the creation of a digital single market and developing e-commerce as strategy to growth and a manner to abolish distances and enables direct between producers and consumers (see European 2020 strategy for growth), many diverse industries are populated by businesses that operate in "two-sided markets or platforms" and serve distinct groups of customers who need each other in some way, providing a common (real or virtual) meeting place and facilitating interaction between members of the two distinct customers groups: providers on one side of the market, and end-users on the other side of the other side.

REFERENCES

Enthoven, A.C., Van de Ven, W.P.M.M., 2007. Going Dutch—managed competition health insurance in the Netherlands. N. Engl. J. Med. 357 (24), 2421–2423.

Europe 2020 – Europe's growth strategy: <http://eurlex.europa.eu/LexUriServ/LexUriServ.do?uri=COM: 2010:2020:FIN:EN:PDF>.

Helderman, Van der Grinten, 2005. Market-oriented health care reforms and policy learning in the Netherlands. J. Health Polit. Policy Law. 30 (1–2), 189–209.

ACKNOWLEDGMENTS

I would like to thank my colleagues and coauthors for their collaboration and support. Special thanks to Tobias Mettler for his collaboration, comments, and time. I also gratefully acknowledge the editorial managers for their efforts to design and assemble this book.

THE PURPOSE AND AIM OF THIS BOOK

This book deals with two-sided and multisided markets phenomena in the e-health area and focuses on the necessity of creating an ecosystem and corresponding strategies that sustain a type of market in which services are delivered online, across distinct organizational or even national borders.

The chapters of this book discuss issues of relevance concerning the e-health market of today such as: (1) how to develop innovative and cost-effective implementation strategies for complex organizations in order to handle the dynamics of two-sided markets, (2) the importance of barriers and facilitators for two-sided markets when implementing e-health services and/or IT-based innovations, and their impacts on business models, and (3) pre-requisites to be achieved in health and social care organizations, which act in two-sided markets, when implementing e-services and/or IT-based innovations.

Based on the introductory remarks, in this book we emphasize the need for implementation models, business models, innovative design strategies, as well as security and usability models that contribute to the emergence of a sustainable and viable two-sided market in the area of e-health. Our emphasis does not imply any presumption on anti-competitive conduct, price strategies, market power, or tools of competition analysis or price-fixing strategies. Our analysis does not imply either that, given our current state of knowledge, competition authorities should be more willing or reluctant to intervene in a two-sided market in e-health as compared to traditional markets.

The knowledge produced and the examples presented in this book can be of importance for decision makers, managers, stakeholders, and other interested individuals that need to plan the transition from piloting e-health services to actual implementation and development of strategies that facilitate the adoption and diffusion of innovative e-health services and/or IT-based innovations in a two-sided market environment. Further, in this book we discuss strategies that stakeholders, vendors, and customers should consider when defining public or regulatory policies as well as when designing and implementing business models with a view to stimulating economic growth of a two-sided market for the e-health area. In view of our aims, this book includes the chapters as listed in the Contents.

PART I

Introduction to the Ecosystem for Two-Sided Markets, Barriers and Facilitators

CHAPTER 1

Introduction to the Ecosystem for Two-Sided Markets, Barriers and Facilitators

V. Vimarlund[1] and T. Mettler[2]
[1]Linköping University, Linköping, Sweden; Jönköping University, Jönköping, Sweden
[2]University of Lausanne, Lausanne, Switzerland

INTRODUCTION

Currently most consumer markets are push markets. Producers and suppliers develop and mass-produce products and then persuade consumers that they need these products, through marketing and commercialization efforts. These push markets have become so extreme that most producers spend upwards of 90% of their capital in marketing and commercialization, leaving very little for production and innovation. This means that new producers cannot break into the market without the necessary marketing infrastructure, as new ideas are difficult to launch without expensive commercialization efforts. Given the costliness of marketing, most producers also compete for the largest market, ignoring marginal products and marginal needs. Consequently, specialized products and services increase cost structures.

In contrast, a pull market begins with the demand rather than the supply. That is to say, the need to produce specific e-health services that fulfill the prerequisites consumers and healthcare providers stated, either by regulation or by managerial decisions. This greatly reduces the need for marketing and commercialization. Producers respond to demands expressed by consumers directly. This removes the barriers to market entry experienced by new producers, developers, and emerging economies. Consumers also drive the design and can thereby steer production in more diverse directions. A pull market is, however, dependent on effective communication, consumer demands to appropriate producers or suppliers, innovative business models, sustainable implementation models, as well as inclusive design approaches, security, and evaluation of outputs.

As e-health undergoes the digital transformation that has occurred in many other sectors of modern society and the global economy (Barrett et al., 2015) moving from a push market to a pull market, it has implied even an evolution of the market from a one-sided market to a two-sided market with all its implications and opportunities.

Two-sided market key characteristics are to serve two distinct user groups that generate value for each other in symbiosis (Eisenmann et al., 2009). Examples of such markets include credit cards, composed of cardholders and merchants; HMOs (patients and doctors); operating systems (end-users and developers); travel reservation services (travelers and airlines); yellow pages (advertisers and consumers); video game consoles (gamers and game developers); and communication networks, such as the Internet. In e-health, a two-sided market is, however, expected in addition to serve distinct groups of users, to benefit providers and consumers of goods and services through brokers that sustain two-sided market phenomena and also to support more coordinated healthcare management because of the emergence of population health platforms in data analytics services over aggregated electronic patient health records providing insights that will support more coordinated healthcare management (Vincent, 2014).

The theoretical literature on two-sided markets is relatively new. Most of the studies of the prerequisites for sustainability of two-sided markets have been performed using industrial examples which highlight issues related to the economic relationship among the various industry participants, as well as the nature of and effects from the interplay between service providers and service consumers. The resulting network effects have been pointed out as the most relevant factor for the sustainability of such a market structure.

A number of papers in the "multisidedness" literature have discussed the difference between one-sided and two-sided market structures. Three types of definitions generally appear in the literature. The first is the one of Rochet and Tirole (2006). "A market is two-sided if the platform can affect the volume of transactions by charging more to one side of the market and reducing the price paid by the other in an equal amount." Comparing this definition with a one-sided market, it is possible to appreciate two main differences: (1) the price structure matters for participating in a two-sided market, and (2) there are membership externalities that do not exist in one-sided markets because of the possibility to determine price by bargaining or monopoly.

A second definition of multisided market is the one proposed by Schmalensee and Evans (2007). According to their definition, "a multisided market has two or more groups of consumers, who need each other, who cannot capture the value of their mutual attraction, and who rely on a catalyst to facilitate their interaction."

A third definition finds that there is a two-sided market when there is "some kind of interdependence or externality between groups of agents that are served by an intermediary" (Rysman, 2009).

In the e-health market, the expansion of the Internet economy, spurred by lower communication costs, reductions in the cost of computing, and software technology advances, has shown to be significant enough to stimulate the entry of business organizations many of which are already in two-sided markets. The relationship among the various industry participants and the existence of two groups of customers (providers of e-health services and customers) cannot, however, solve externalities by themselves and in many cases there is a need for some kind of regulation, as in the case of Scandinavian countries where the prices the consumers pay are regulated by policies or other price strategies (subvention, differentiated pricing due to maximum roof costs, etc.).

THE ECOSYSTEM OF A TWO-SIDED MARKET IN E-HEALTH

The fundamental role of a two-sided ecosystem for a two-sided market is to enable parties to realize gains from trade or other interactions by reducing the transactions costs of finding each other and interacting. Two-sided markets do this by matchmaking audience, building audiences, and minimizing costs for duplication, advertising supported media, building audiences and exchanges.

A market is characterized as being able to facilitate exchanges between two groups who can generally be considered "buyers" and "sellers." The exchange[1] helps buyers and sellers search for feasible contracts—that is where the buyer and seller could enter into a mutually advantageous trade—and for the best prices—that is where the buyer is paying as little as possible and the seller receiving as much as possible.

Exchanges provide participants with the ability to search over participants on the other side, and the opportunity to consummate matches. For instance, health and/or social care, personnel search for available alternatives of services. Having large numbers of participants on both sides increases the probability that participants will find a match. However, a larger number of participants can lead to congestion if no variation of providers of similar or identical services exists.

In some cases, some exchanges charge only one side. Healthcare organizations pay for the services delivered and used by some groups of patients, while patients get them for free or pay a minor amount of money for them. In other cases, exchanges charge both sides. Insurance brokers historically charged both insurance customers and insurance providers in some types of transactions.

Further, in e-health, a two-sided market is composed of different stakeholders, who act in parallel. For example, public agencies or public organisations, decision makers (e.g. politicians or other governmental stakeholders), patients and/or consumers of services (citizens), and practitioners (experts that produce and deliver services) and that represent:

- *Service consumers* use the services and define the usage goals for the services, i.e. tasks that need support. They may also report on problems and failures in the service usage and provide feedback for the service validation.
- *Service providers* are independent members that provide digital services to be used by other members or consumer participants in the two-sided platform.
- *Service brokers* promote the services, enable service delivery, and match the demand with the best available services.
- *Infrastructure providers* render services that implement the purpose and capabilities of the platform, such as establishing, modifying, monitoring, and terminating collaborations, and operations for joining and leaving collaborations.

The *business ecosystem* may emerge spontaneously due to a common interest or demand, or as a result of long-term strategic planning. The members share the common regulation of the area but are able to act independently, and join and leave the ecosystem freely, since there is no dependency between members. In some cases, organizations, firms, and companies work together for the delivery of e-health services because the possibility of combining knowledge and money creates additional value in the network of actors.

An ecosystem for a two-sided e-health market has further to be able to support parties to realize any form of gains from interactions and trade with different actors by reducing the transaction costs or duplications costs for finding and dealing with each other. A first prerequisite is to have a platform (i.e., an e-health portal) that attracts groups to view, participate, share, and search for matching between demand and supply or e-health services. A second prerequisite is the coexistence of an ecosystem, which is composed of a business ecosystem, and a digital infrastructure ecosystem:

- *The business ecosystem* provides information about self-regulation issues, network alternatives, rules, taxes, and all types of economic information that could affect the development of business and reimbursement models, but also information about the existence of brokers that can influence or make decisions about the offerings to the customers of a health and social care sector, information about agents (providers and consumers) that will consume some specific service, and information about the number of alternative compatible agents on the opposite market side that are capable of offering an e-health service and even information about compatibility, similarity, and use of business model alternatives available.
- *The digital infrastructure ecosystem* exists in parallel with the business ecosystem and provides information about the socio-technical complexity of the system (the platform) and consists of a combination of carbon-based servers (humans) and silicon-based servers or any other forms of automated infrastructure that enables service providers sharing the service taxonomy, service description (categorized by domain purpose and technology). However, the digital infrastructure also includes information about the kinds of services, technology, and consumers that interact and utilize the delivered services. The role of the digital infrastructure ecosystem is, consequently, to focus on technical, syntactic and semantic interoperability, safety, and security issues, as well as in the interactions between systems and humans, and its dependency between ecosystem members.
- *Furthermore*, a digital infrastructure in a two-sided e-health market demands the existence of a knowledge management module that enables reuse of existing best practices to ensure that quality requirements and quality driven methods are used, and that information about the repository unit for storage of the collaboration models, ontologies of service types to support interoperability, validation to guarantee the effectiveness of the service ecosystem is performed, and semantic interoperability and alignment among ecosystem members, services, and technologies is maximized.

CONSTRAINTS, THREATS, AND EXCEPTIONS THAT AFFECT THE SUSTAINABILITY AND SIZE OF THE E-HEALTH TWO-SIDED MARKET ECOSYSTEM

Two-sided markets are dynamic and evolve all the time. They need rules and regulations because the behavior of their members can affect the value of the two-sided platform as a whole.

A two-sided ecosystem in e-health should be able to describe the model that defines the properties and infrastructure of the ecosystem, and how they are implemented. Such a model describes, for example, how a trusted collaboration can be established between members, interaction rules and regulations, processes for how to join and leave the ecosystem, how to describe and deliver services, how knowledge will be managed, and how trusted collaboration can be established between members of the ecosystem.

Since a two-sided market is dynamic, it evolves all the time as new members, services, and value networks emerge. Consequently, new support services should emerge or evolve in parallel, thus allowing service providers to detect easily what kind of services have demand inside ecosystem. In addition, as the ecosystem monitors the quality of its services, it should also provide a matchmaking service for service selection to match the required quality with the provided quality.

In the case of e-health service demand, there is no embedded knowledge in business experts about what the real needs for different groups are. For this reason, a two-sided ecosystem should be based on interactive principles, and allow individuals to influence the supply and development of novel services using, e.g., some kind of interaction that allows consumers to influence decisions for critical services (e.g., up or down voting), through participation in the service identification (e.g., open innovation challenges), and specification (e.g., crowdsourcing documentation). Furthermore, participants from at least two of the four pillars of a society (e.g., government or international institutions, corporations, and business/enterprises or services organizations, including NGOs and individual citizens) have to be persuaded to change the paradigm of the market for e-health services.

An important issue is, however, to be able to restrict the ecosystem to "good things" that actually help solve problems related to market inefficiencies (e.g., monopolies, cartel, negative external effects) and not to restrict the development of the ecosystem to create global public goods, but rather to stimulate economic growth and global cooperation and competition. For this reason, producers and consumers of e-health services must overcome the challenges that the market presents today and minimize issues that affect participation, transaction, and market power unilaterally or through coordinated actions with other firms, or difficulties to entry because of the presence of particularly strong inter-group network effects or because of the existence of legislation and rules not easy to change, and consequently the sustainability of the market.

Five fundamental factors determine the size and sustainability of two-sided markets in the e-health area: (1) Ownership or its institutional arrangement nature of e-health services, (2) different kinds of network externalities, (3) price models, (4) adhesion pattern (multihoming vs. single-homing), and (5) the chicken/egg problem. We will expand on these points below.

Ownership and/or its institutional arrangements in the nature of e-health services

One necessary piece of information for attracting firms and organizations willing to produce and offer innovative e-health services in a two-sided platform is to indicate if the services may be owned by an intermediary (i.e., a county council, a healthcare unit, an insurance company, a pharmaceutical company) or by agents active on each side of the ecosystem (i.e., a patient organization, or the company that produce the service), or if another ownership model based on a series of contractual arrangements and property rights for the services offered will be developed. In such circumstances it will be necessary to identify who has the right to restrict entry, access, and use of the services and if there will be some restrictions to people or organizations that are not included in the arrangements, and which contractual arrangements and property rights will be developed and regulated (as in the case of the virtual patient record, the accessibility of which is restricted to the individual patient and healthcare professionals). Types of ownership will further influence the business ecosystem. For this reason it is important to identify if the business ecosystem will be:

- Coincident and offering products or services on the same sides. That is the case in games for special treatments, telemedicine systems, and payment cards.
- Intersecting: offering products or services that are substitutable at less on one side, as in the case of security systems that do not support solutions that are not linked with the ones they offer to their customers.
- Monopoly with no competition on any side. Although this could exist in theory, of course, it is hard to identify any industry for which this alternative could be sustainable today (yellow pages was an example for a time perhaps in some places).

Different kinds of network externalities

Network externalities are said to exist when consumer utility in a certain market depends (usually in a positive way) on consumption of the same goods or service by other agents, and is affected by the specific price applied to that side. They, however, do not depend on consumption of agents in the same class (e.g., consumers of the same product), but on consumption of different, but "compatible," agents on an opposite side of the ecosystem.

For example, in joining an intermediation (or exchange) service ecosystem, as a telemedicine service, a buyer will take into account the number of potential sellers using the same service. However, even when the healthcare system is supported by taxes or other kinds of regulations and policies, the effects on the price the organization pay will influence the price pattern across both sides of the market. It is, therefore, possible that the price structure to get both sides on board and optimize the usage of the services becomes asymmetric because of the price of the service for one side being above the one for the other side. Some alternative models are discussed in the next paragraph.

Price models

The crucial difference between pricing instruments in business ecosystems, compared with a single market, is that it is not necessary to have all the sides "on board" controlled by regulations or policies. The price structure used will much depend on the cross-price elasticity mechanism to subsidize groups of users, or the use of price instruments chosen depending on the range of options applicable. For instance:

- *Lump-sum* basis, that is, a tariff that does not explicitly depend on how well the ecosystem performs. One example is Windows OS, which is generally sold at a posted price.
- *Tariff as function of the performance* is the opposite from the above mentioned mechanism. One example of this practice is a TV channel or a newspaper that makes its advertising charges an increasing function of the audience or readership it obtains (to do this there must be a credible third party which can accurately estimate audiences).
- *Actual interactions* or signing credible contingent contracts making payments dependent on subsequent participation and transaction levels. Complicated contracts obviously have the potential to extract consumer surplus more fully, but in some circumstances could also make a dominant firm much more susceptible to entry and thus greatly limit profits. For instance, a potential intermediary could attract all buyers by promising to make large payments to them if it fails also to attract all sellers away from the incumbent intermediary.

In a two-sided e-health market it is rational to expect that organizations or producers of services subsidize consumers or choose some beneficiaries, and charge developers, i.e., with licenses or accreditation cost.[2] This is because of policies, regulations on both sides, or because of welfare issues. Price models in two-sided e-health markets are expected to do be influenced by externalities, price control, and even price discrimination, absence of substitutes and asymmetric information, absence of new entrants, in some cases on both sides.[3]

Adhesion pattern (multihoming vs. single-homing)

Whenever there are several providers of the same types of services or products, customers on each side of the market may choose to subscribe to one provider only ("*single-homing*") or to several providers ("*multihoming*"). The concept of multihoming covers both subscribers to all available providers ("full" multihoming) and to more than one (but not all) of them—partial multihoming (clearly this distinction does not arise where there is a monopoly solution).

Multihoming can be more easily observed when fixed costs for accessing the service are low or absent. On the contrary, if consumers pay only a fixed subscription fee for a service, they will tend to use a single supplier, especially if the supplier offers comparable services and has similar degrees of acceptance among consumers.

Multihoming makes, however, the whole analysis of the sustainability of the market considerably more complex, and demands the adoption of different subscription policies both within and across the sides of the market, depending on preferences and possible differentiation among providers' offers. It is important to note that the presence of multihoming influences the degree of competition (Rysman, 2007) and avoids situations such as the ones that appear in single market alternatives (i.e., monopolies or quasi monopolies) that usually end in high prices and inefficacies for the social welfare system.

Multiside alternatives are, however, based on a series of assumptions that have to be fulfilled. They are: (1) that there is no differentiation among different providers in the sense that they are firms or organizations that can deliver the same service and fulfill the quality issues the area demand, (2) that customer preferences are sufficiently homogeneous, and (3) that customers on the multihoming side have no bargaining power that allows them to limit rent extraction by the provider.

Chicken/egg problem

Which came first? This causality problem is particularly relevant for markets that depend on network externalities such as e-health. Who would buy the first phone if there is no one to call?

Innovative e-health services are often of little value initially, because there are no actors to mediate between. Thus, a new e-health service typically follows distinct life cycle phases of rollout and operation. The first part is the most difficult one, because without having an initial installed ecosystem, there are no possibilities to diffuse innovations due to the small amount of consumers. Although each case is different, there are some generic strategies to overcome this problem:

- Developing cooperation and alliances that include customers, suppliers, rivals, and information about the market complementary products and services (Shapiro and Variant,1999).
- *Expectation management*: Expectations can easily become a self-fulfilling prophecy in markets with positive feedback. To manage expectations, you should engage in aggressive marketing, make early announcements of new products, assemble allies, and make visible commitments to your technology.

- *Penetration pricing*: To challenge established brand loyalty and lock-in, one can appeal to price-conscious consumers with a low initial price or "give-away strategies" (Stabell and Fjeldstad, 1998). This strategy can also have internal benefits. With a domestic pressure to control costs, the company may develop greater efficiency and thereby institutionalize it as a competitive advantage.

- *Be good*: Consumers, as is widely known, let their emotions play a major role when making buying decisions. Correspondingly, companies are increasingly letting corporate social responsibility (CSR) policies propagate into their procurement processes (e.g., environmentally friendly products). Distinguishing oneself positively on such "soft issues" may therefore pose a direct competitive advantage and benefit the brand.

- *Handle lock-in*: If there already is a dominant player in the market, there are several strategies for gaining foothold (Johnson, G., Scholes, K., and Whittington, R). For example, leapfrogging. Instead of launching an imitation service, launch a superior alternative. This may be of particular relevance in markets with rapid technological advances, which big players tend to handle conservatively (i.e., structural inertia and longer life-cycles; focusing on robustness rather than on the technological frontier).

- *Low initial access barriers*: As long as other switching costs are relatively low, the required effort to join can be decisive. Therefore, the process of joining a network should be as effortless as possible. In fact, if easy enough to join, there might be no need of switching at all: Customers may join your network while keeping another foot in a competing one (e.g., many PC-users use both MSN and Skype, and buy from more than one online retailer). One should also consider "viral" recruitment of new members, which Facebook has employed successfully (with a few clicks, Facebook can send auto-generated invitations to your contacts, provided that you share login information to services like MSN, Yahoo, or Gmail, or upload your contacts as a file). Likewise, one can develop strategies that stimulate viral recruitment indirectly (e.g., mobile subscriptions that allow you to call family and friends at lower rates).

- *Active recruitment of strategic customers*: In order to get the bandwagon rolling, it might be necessary to invest in active recruitment of strategic customers. This is particularly important if there are customer groups that depend on one another (e.g., retailers and consumers). It may be sensible to build up a critical mass among one customer group, before attending another (i.e., retailers before consumers). To make this process work, it might be required to make concessions (limited exclusivity, lifetime access to premium services). Although often perceived as something to avoid, one can go as far as offering shares to catch a "really big fish" (which also provides incentives to the customer to expand the network proactively).

- *Sweeteners*: Besides penetration pricing and low initial access barriers, one should consider offering additional sweeteners on top. This could be in the form of gift cards to be used within the network (thereby also triggering activity within the network), beneficial access terms when joining before a given date, or "welcome gifts" (these days USB sticks seem to be in fashion). If possible, one should make use of "cooperative sweeteners" (e.g., alliance partners willing to provide a limited set of items for free, or at a discounted price). The latter requires the cooperative party to see benefits from it, such as branding and expansion of the customer base.

- *Making customers come back*: Once a small installed base is in place, it is important to stimulate activity in the network (it is less the numeric customer base that is of value than the actual use of the network). One strategy is to build loyalty (like providing convertible bonus points for purchases), another (complementary) is to keep the "news factor" alive (such as optional newsletters containing special offers and announcements). This, of course, should not be overdone. Excessive newsletters can be perceived of as spam, and loyalty schemes are commonly deemed bothersome.
- *Timing and chronology*: Providing access to a premature service may result in negative attention and scare potential customers or partners away. Likewise, engaging in expectation managing before an initial "go live" can spoil the surprise effect, thereby providing the competition a free chance to develop preemptive strategies. There are no definite recipes on timing and chronology—a lot depends on the market conditions at hand. Therefore, it is important to know particularities of the market and develop contingency plans to be prepared for eventualities. Employing a bit of game theory may also be very helpful.

THE IMPORTANCE OF THE EXISTENCE OF NETWORKS

For two-sided markets it can be important to recognize that competition on both sides of a transaction can limit profits, especially in situations where multihoming is nonexisting. Ecosystems with more customers in each group are consequently more valuable to the other group and vice versa. Further, price equaling marginal costs, such as the ones existing in healthcare organizations, is not a relevant economic benchmark for many firms; even when in the area of e-health it is not possible to conclude that deviations between price and marginal costs on one side provide any indication of market power or sustainability. It is therefore important to identify dimensions of competition and focus on sources of inefficiency or constraints that can influence incentive for being a part of the ecosystem.

An e-health two-sided market ecosystem, by definition, tries to achieve a change or to provide an infrastructure where other stakeholders sell, organize, and offer different collaboration between countries on the same hierarchical level (horizontal collaboration) but at the same time with the possibility to offer both public funded, private funded, and micro financed services and goods. The e-health area demands the presence of different networks to address problems or to solve trans-boundary, cross border, or even simultaneous problems to ensure its sustainability:

- *Knowledge networks* to develop new thinking, research, ideas, and policies, or to discuss security and safety issues that can be helpful in solving transnational problems. Their emphasis is on the creation of new ideas, not their advocacy. Knowledge networks should help to scale continuously new collaborative ideas, encourage participation from individuals, organizations, and companies no matter where they live or what language they speak. The knowledge network in an e-health ecosystem should support knowledge creation around products, services, and goods, and disseminate ideas for open innovation and worth spreading in order to map developers' and customers' needs in a more realistic manner.

- *Policy networks* to create government policy even though they are not networks of government policy makers. They may or may not be created, encouraged, or even opposed by formal governments of government institutions. However, powered by global multistakeholder collaboration they are becoming a material force to be reckoned with in global policy development. Their activities cover the range of steps in the policy process, beyond to policy proposals, including agenda setting, policy formulation, rulemaking, coordination, implementation, and evaluation and developing of security policies and principles. Their expertise can often play an important role in global debates and the establishing of norms.

- *Advocacy networks* to change the agenda or policies of governments, corporations or other institutions. An advocacy network challenges business leaders to rethink not only their business strategy, but also their larger purpose and role in the global marketplace. Such a network should contribute to visualizing issues of importance for the e-health area and ask for direct action, lobbying governments and ensuring that the views and values of relevance for the e-health market influence the decisions that affect them.

- *Watchdog networks* to ensure the business of transparency, scrutinizing institutions to ensure that they behave appropriately. Topics range from human rights, corruption, and the environment, to financial services. Customers and citizens can evaluate the worth of products and services at levels not possible before. Stakeholders participating in an e-health ecosystem should characterized for being transparent, prepared for developing a brand or reputation that is sustainable over time, and facilitate management systems, early warning systems, or visualization of open source support systems.

CONCLUSION

An e-health two-sided market needs to offer structures with fixed components, independent of the number of transactions in the form of subscriptions, feed, honor cards, special rules contractually defined, or the possibility to discriminate among agents of the opposite market side. It should contain mechanisms for identifying who pays what, if any payment is needed, in order to get both sides on board. It should also hold constant the total of the prices faced by all the parties. This is because any change in the price structure (or distribution of the goods and services) would affect participation, and reduce the number of interactions.

A two-sided e-health market ecosystem needs to support a flexible model that must meet unexpected demand and therefore be able to handle high demand peaks and long periods of low workload. We have discussed constraining factors, such as the ownership or its institutional arrangement nature of e-health services, different kinds of network externalities, price models, adhesion pattern (multihoming vs. single-homing), and the chicken/egg problem.

Hence, e-health market ecosystems, first and foremost, need to be designed for adaptability and network effects. We have discussed different forms of networks that need to be in place, such as knowledge networks, policy networks, advocacy networks, or watchdog networks. While there is no guarantee for success, this may realize gains from trade or other interactions by reducing the transactions costs of finding each other and interacting.

ENDNOTES

1. The term "exchange" covers various matchmaking activities, for instance, services offered by employment agencies, auction houses, internet sites for business-to-business, person-to-business, and person-to-person transactions, in which various kinds of brokers (insurance and real estate) and financial exchanges for securities and futures contracts are needed.
2. Markets are assumed to be effective when prices direct consumers and firms to behave efficiently, as a consequence of the intrinsic agreement that develops between sellers and buyers, at the moment their trade is in an open single market. Further, any shift in the supply or demand of a product or service affects the price and creates trade barriers, bringing some form of inefficiency.
3. Competition applies pressure to the sides of the see-saw, stretching the rubber band and reducing prices; but it may also shift the balance of the see-saw, so that the direction of price effects is unclear. Similarly, welfare considerations in two-sided markets are complicated by externalities across the two sides.

REFERENCES

Barrett, M., Davidson, E., Prahbu, J., Vargo, S.L., 2015. Service innovation in the digital age: key contributions and future directions. MIS Q. 39, 135–154. 2015.
Eisenmann, T., Parker, G., Van Alstyne, M., 2009. Opening platforms: how, when and why? In Platforms. In: Gawer, A. (Ed.), Markets and Innovation. Edward Elgar, Cheltenham, UK.
Johnson, G., Scholes, K., Whittington, R., 2005. Exploring Corporate Strategy, 7th ed. Prentice Hall, Harlow.
Rochet, J.C., Tirole, J., 2006. Two-sided markets: a progress report. The RAND J. Econ. 37 (3), 645–667. September Version of Record online: 28 JUN 2008.

Rysman, M., 2007. An empirical analysis of payment card usage. J. Ind. Econ. 55, 1–36.

Rysman, M., 2009. The economics of two-sided markets. J. Econ. Perspect. 23, 125–144.

Schmalensee, R., Evans, D.S., 2007. The industrial organization of two-sided platforms. Competition Policy Int. 3 (1), 151–179.

Shapiro, C., Variant, H.R., 1999. Information Rules. Harvard Business School Press, Boston, MA.

Stabell, C.B., Fjeldstad, Ø.D., 1998. Configuring value for competitive advantage: on chains, shops, and networks. Strateg. Manage. J. 19 (5), 413–437. http://dx.doi.org/10.1111/j.1756-2171.2006.tb00036.

Vincent W, Using technology to optimize population health care coordination outcomes. Health Care Informatics. 2014. <http://www.healthcare-informatics.com/article/using-technology>.

PART II

Usability and Design

CHAPTER 2

Patient Safety and Health Information Technology: Platforms That Led to the Emergence of a Two-Sided Market

E.M. Borycki and A.W. Kushniruk
University of Victoria, Victoria, BC, Canada

INTRODUCTION

Over the past several decades we have seen the modernization of healthcare with the introduction of health information technologies (HIT) such as the electronic health record (EHR) (Ashish et al., 2009), mobile E-health applications, peripheral devices used in conjunction with mobile phones (e.g., blood sugar monitors, oximeters) and wearable devices that enable remote monitoring of patient physical activity (e.g., number of steps taken, quality of sleep), cognitive status (e.g., mood), and physiologic status (e.g., blood pressure, respiration, heart rate) (Househ et al., 2012). Even as these technologies have improved the quality and safety of healthcare (i.e., reduced the number of medical errors made by health professionals and patients) while improving monitoring of potential health issues, new HIT has also introduced a new type of error (i.e., technology-induced error) (Borycki & Kushniruk, 2008; Househ et al., 2012; Kushniruk et al., 2005). The aim of this book chapter is to: (1) review the current state of patient safety and HIT in relation to a two-sided market, (2) describe two technology platforms that have aided in the development of a two-sided market in identifying, describing, and reporting on the incidence of technology-induced errors, and (3) discuss future directions involving technology platforms that will continue to facilitate development of this two-sided market that improves vendor and healthcare organizational production of high quality and safe HIT for healthcare consumers and health professionals. In addition to this, the book chapter will detail how two technology platforms have stimulated the development of a two-sided market focusing on HIT safety by disseminating information about HIT safety to healthcare consumers, health professionals, healthcare organizations, and governments, while creating new public policy and the development of approaches and methods improving the quality and safety of HIT internationally. The authors begin by defining technology-induced errors.

TECHNOLOGY-INDUCED ERRORS: A DEFINITION

Technology-induced errors can be defined as those sources of error that "arise from: (1) the design and development of technology, (2) the implementation and customization of a technology, (3) the interactions between the operation of a technology and the new work processes that arise from a technology's use" (Kushniruk et al., 2005), and (4) the exchange of information between two differing technologies (Kushniruk, Surich & Borycki, 2012). Technology-induced errors may lead to patient harm, disability, or death (Magrabi et al., 2010; Magrabi et al., 2012). Technology-induced errors represent a new type of medical error introduced by HIT or where HIT contributed to patient harm or death (Borycki & Kushniruk, 2005; Borycki & Kushniruk, 2008). Technology-induced errors have become an international concern for healthcare consumers, health professionals, and healthcare organizations who use HIT and design/develop and implement it to support patient health promotion and/or self-management of chronic illnesses (Borycki & Kushniruk, 2008; Kushniruk et al., 2013).

THE COST OF TECHNOLOGY-INDUCED ERRORS

Technology-induced errors are a significant concern for patients, health professionals, healthcare organizations, and vendors. For patients, there is the pain and suffering associated with physical harm to the individual (Magrabi et al. 2010, 2012). For health professionals, who are involved in a medical error, such as a technology-induced error, this can lead to emotional and job-related distress (Waterman et al., 2007). For healthcare organizations (e.g., hospitals, clinics) and for vendors, the financial costs can be significant. Costs of are two-fold for healthcare organizations and vendors: (1) the costs associated with re-design, re-programming, and re-implementation, (2) the costs associated with medical care of harmed patients and lawsuits. Therefore, identifying and fixing a technology-induced error prior to implementation is important (Borycki & Keay, 2010).

Information about technology-induced errors is important as it may be needed to initiate software fixes to existing and implemented HIT (Borycki & Keay, 2010; Horsky et al., 2005). Improving the safety of HIT early in the software development lifecycle, prior to implementation, would lead to cost savings for vendors (Borycki & Keay, 2010). Vendors that design, develop, and deploy safe software would have the opportunity to improve the safety of their software and reduce the costs associated with a software's re-design, re-programming, and re-implementation when technology-induced error prone software features and functions are discovered after implementation (Borycki & Kushniruk, 2008; Borycki & Keay, 2010). According to Rothman (2000), each software fix post implementation costs up to $500,000 to correct. Fixes initiated earlier in the software development lifecycle, prior to systems release, cost up to $50,000—a 90% cost savings when compared to fixes that take place after implementation. Healthcare organizations and vendors benefit from such cost savings by attending to technology-induced error events after HIT has been implemented (Borycki et al., 2009). Baylis and colleagues (2011) identified that healthcare organizations and vendors could achieve a 36.5–78.5% cost savings if strategies for testing for technology-induced errors were integrated early into the software development lifecycle. In their work they identified how a portion of these cost savings could be achieved by avoiding costs associated with treating patients who were harmed and lawsuits involving technology-induced errors.

Health professionals and healthcare consumers would welcome changes that would improve patient safety where HIT is concerned (Boulos et al., 2014). It is argued that consumers would be more likely to purchase software that would improve safety (much as consumers who are concerned about safety are more likely to purchase an automobile that has a good safety record during car crash tests). Consumers, healthcare organizations, and vendors would also want such improvements in the safety of HIT so there would be a decrease in the occurrence of technology-induced errors as this would lead to an overall reduction in costs associated with re-design, re-programming, and re-implementation of HIT and the costs associated with lawsuits where technology may be a contributing factor in patient harm or death (Baker, 2011).

THE EMERGENCE OF PLATFORMS TO ADDRESS TECHNOLOGY-INDUCED ERRORS

Reports of technology-induced errors have increased over time (Magrabi et al., 2012; Palojoki et al., 2016a,b; Samaranayake et al., 2012). Two platforms have played a key role in disseminating information about technology-induced errors: Medline® and MAUDE®.

PLATFORMS FOR ADDRESSING TECHNOLOGY-INDUCED ERRORS

There exist a number of platforms for disseminating information about technology-induced errors to healthcare consumers, health professionals, healthcare organizations, and vendors, yet two platforms [i.e., Medline® (see—http://www.ncbi.nlm.nih.gov/pubmed) and MAUDE® (see—https://www.accessdata.fda.gov/scripts/cdrh/cfdocs/cfmaude/search.cfm)] have emerged has having a critical role in the advancing a two-sided market involving HIT with a focus on improving HIT safety. These platforms have documented the presence of technology-induced errors, allowed for some description of technology-induced errors, and provided information about their incidence. Medline®, one platform, is a searchable online database that can be searched anywhere and anytime. MAUDE® is a publically available incident reporting system for medical devices. Medline® articles and MAUDE® data are publically available to healthcare consumers and health professionals. Medline® articles and MAUDE® data have been analyzed by researchers and have been used to develop strategies to address and disseminate information about technology-induced error management. In the next section of this book chapter we will discuss Medline® and MAUDE® as platforms for fostering a two-sided market aimed at improving the quality and safety of HIT with consumers and health professionals on one side of the market and healthcare organizations and vendors on the other side.

MEDLINE®

"MEDLINE (Medical Literature Analysis and Retrieval System Online, or MEDLARS Online) is a bibliographic database of life sciences and biomedical information. It includes bibliographic information for articles from academic journals covering medicine, nursing, pharmacy, dentistry, veterinary medicine, and healthcare" (https://en.wikipedia.org/wiki/MEDLINE). Medline® is used by health professionals and healthcare consumers to learn about health and disease and its treatment. More recently, Medline® has also become one of the most robust sources of information about HIT used in healthcare as it remains the main searchable database that indexes health and biomedical informatics articles published in journals (Medline®, http://www.ncbi.nlm.nih.gov/pubmed).

Medline® has played an important role in addressing technology-induced errors. Initially, Medline® indexed articles describing technology-induced errors. More recently, it has become a significant source of information (in the form of research and reviews) about how HIT can be designed, tested, customized, implemented, and regulated so that HIT can become safer over time. Early studies documenting the presence of technology-induced errors were first published in disparate health journals. Medline® librarians indexed these articles and provided a single database where published research about technology-induced errors could be further reviewed and reported on in scoping and systematic reviews (e.g., Borycki & Keay, 2010, Borycki et al., 2012a,b).

To illustrate, early publications about technology-induced errors began to emerge in 2005. For example, an early publication by Koppel and colleagues (2005) documented the presence of technology-induced errors such as "separation of functions that facilitate double dosing and incompatible orders, and inflexible ordering formats generating wrong orders" (Koppel et al., 2005, p. 293). In their work the researchers conducted interviews, focus groups, observations, and surveys with physicians, nurses, pharmacists, and healthcare administrators at a healthcare organization that had implemented a computerized order entry system in the United States. The researchers observed and learned from participants that HIT could facilitate technology-induced errors (Koppel et al., 2005). In another study published in the same year by Kushniruk et al. (2005), the researchers were able to document a relationship between poor HIT usability and technology-induced errors. In their publication the researchers identified specific features and functions of a mobile e-prescribing system that led to technology-induced errors (e.g., default medication doses). In some cases physician participants were able to correct the technology-induced error (having observed it). In other cases these errors were missed and led to medication errors such as the wrong dose of a medication being prescribed (Kushniruk et al., 2005). These publications and other technology-induced error publications have been indexed in the Medline® database.

This has helped healthcare consumers and health professionals who use this database to learn about the latest updates from a healthcare research perspective. Health professionals and consumers have accessed these articles and others, learned about technology-induced errors, and in turn have reported on technology-induced errors in medical device incident reporting systems (Magrabi et al., 2010, 2012). Publications indexed in Medline® have also led other researchers around the world to study HIT safety and to develop new solutions to this emerging international problem (Borycki et al., 2011; Palojoki et al., 2016b; Samaranayake et al., 2012).

More recently, this research as indexed by Medline® has come to the attention of policy makers internationally in countries such as Australia, Canada, the United States, and the United Kingdom. Policy makers have used the research to inform policy making activities surrounding HIT safety (Kushniruk et al., 2013). For example, in the United States, the Institute of Medicine (IOM) in 2011 published a report on HIT Safety which provides a roadmap going forward about how to tackle the growing issue of technology-induced errors and includes a discussion about regulation of the HIT industry (IOM, 2011). The report outlines a plan for how HIT might be regulated for safety which has been used by the Office of the National Coordinator in the United States to inform HIT-related policy initiatives aimed at improving incident reporting and the safety of HIT (Office of the National Coordinator, https://www.healthit.gov/newsroom/about-onc). In other countries (i.e., the United Kingdom and Australia) initiatives are underway to provide guidelines that support the safe customization, updating, and development of safer HIT (e.g., user interface guidelines such as the Common User Interface developed through a partnerships between the National Health Service in the UK) and Microsoft (NHS and Microsoft, National Health Service and Microsoft n.d. http://systems.hscic.gov.uk/data/cui) and the Australian Practice Guidelines represents an extension of this work. Additionally, these guidelines are now being used in procurement processes (CUI guidelines) (Australian Commission for Quality and Safety in Healthcare, 2012). European Union countries such as France are now employing usability testing as standard practice for medical devices to provide feedback about the quality and safety of the HIT (Kushniruk et al., 2010).

The use of Medline® by healthcare consumers and health professionals as a technology platform as a source of information about technology-induced errors and the methods for reducing them is important. As a powerful database Medline® is used by many health professionals (i.e., doctors, nurses) and HIT professionals (i.e., health informatics) from around the world. Medline® disseminates information about technology-induced errors quickly to varying parts of the world (as it is used internationally by health professionals and researchers as a source of research based information). "On the average day in April 2015, approximately 3.5 million searches were performed on the PubMed Website (pubmed.gov). An additional 5.2 million searches were done by scripts (e.g., by application programming interfaces or APIs)." (Medline, https://www.nlm.nih.gov/services/pubmed_searches.html) Medline® led researchers from around the world to report on the safety of HIT in vary contexts (e.g., Australia, Canada, Finland, United States) and to work on studying how HIT can be made safer. The information found in research publications provided by Medline® was used by policy makers who are now engaged in strategies outlined above (e.g., use of guidelines for procurements) to improve HIT safety internationally. Without Medline® as a platform this would not be possible and a two-sided market would not have been created.

MANUFACTURER AND USER FACILITY DEVICE DATABASE (MAUDE®)

Manufacturer and User Facility Device Database (MAUDE®) is a database that "houses medical device reports submitted to the Food and Drug Administration (FDA) by mandatory reporters (manufacturers, importers, and device user facilities) and voluntary reporters such as healthcare professionals, patients, and consumers." (MAUDE®, https://www.accessdata.fda.gov/scripts/cdrh/cfdocs/cfmaude/search.cfm). A unique aspect of MAUDE® is that it allows health professionals, health informatics professionals, and administrators to report HIT incidents involving technology-induced errors directly to the Food and Drug Administration (FDA). Of note the reporting database is open access and the incident reports can be submitted by healthcare consumers, health professionals, health informatics professionals, healthcare administrators, and vendors. As well, researchers are able to download anonymized, incident reports for review and further analysis (Magrabi et al., 2012). The MAUDE® incident reporting database is a platform that allows for exchange of information about technology-induced errors between healthcare consumers/health professionals and researchers, vendors, healthcare organizations, and governments. MAUDE® is unique as it allows for open access to technology-induced error incident data. Most incident reporting systems do not provide such information to researchers (Mullner, 2009). Further to this, the incident data supports an exchange of information between vendors/healthcare organizations and health professionals/consumers/health informatics professionals about technology-induced errors. It is more typical for such incident data to be collected in databases that are not readily accessible for review and analysis by health informatics safety experts from around the world (Mullner, 2009).

MAUDE® has been used to investigate the state of incident reporting involving technology-induced errors by health informatics researchers. It has also been used to develop an initial classification system for these types of events (Magrabi et al., 2010, 2012). Researchers have identified that many of the technology-induced errors reported by health and health informatics professionals that are present in the MAUDE® database are similar to those arising from analysis of proprietary or not publically accessible incident reporting databases (Samaranayake et al., 2016; Palojoki et al., 2016a,b). Yet, differences remain in the types of incidents involving technology-induced errors that are present in the MAUDE® data and data collected in other jurisdictions in countries such as China and Finland (Samaranayake et al., 2016; Palojoki et al., 2016a,b). Initial analyses using MAUDE® data represent a first attempt to characterize the problem and develop an initial classification scheme of technology-induced error reports. To obtain a robust classification scheme, analyses of incident reports need to take place in other countries with robust, comprehensive incident reporting systems in highly digitized healthcare settings and contexts to fully understand technology-induced error reporting, and over time as new technologies are introduced into the healthcare marketplace.

To illustrate, Magrabi and colleagues (2012) published information about technology-induced errors following an analysis of FDA data downloaded from MAUDE® website. The researchers were able to report on the types of errors that led to near misses, harms, and deaths as well as the incident rate based on MAUDE® reports (Magrabi et al., 2012). This work was later extended to other countries such as the United Kingdom (Magrabi et al., 2015). Other researchers in more highly digitized settings have reviewed their healthcare incident reporting data and have developed alternate coding schemes and typologies (Samaranayake et al., 2012) with a high level of fit between the incident report and coding scheme (Palojoki et al., 2016a,b).

Due to MAUDE®'s online availability as an error reporting system and the availability of anonymized error reports, this information has played an important role in initiating a two-sided market with MAUDE® as the technology platform for exchanging information about HIT safety between healthcare consumers, health professionals, and health informatics professionals and vendors/healthcare organizations. Again these initial works arising from analyses of MAUDE® data have come to the attention of policy makers. Calls have been made for incident reporting data to be more fully analyzed to learn more about how technology-induced errors occur and how this data could be used to inform improvements in HIT safety. Yet, this would not be possible without such publically available data that could be downloaded for review and analysis (i.e., the FDA MAUDE® data being published online and being for used research and analysis of technology-induced errors).

Research involving FDA MAUDE® data has also informed policy maker and regulatory initiatives in the area of medical device safety, definitions surrounding what is or is not a medical device, and an exploration of how medical software such as EHR should be viewed and regulated, in several countries around the world. Such information has informed vendors of newly developed software and has led to the development of new approaches to testing for software safety (that are being used in some parts of the world as a way to identify technology-induced errors) prior to HIT implementation in a healthcare setting such as a hospital (Borycki & Keay, 2010). For example, the Office of the National Coordinator (ONC) is now working toward helping vendors to improve the usability and safety of electronic health records in the United States even as health professionals such as physicians and nurses continue to report on technology safety issues (ONC, n.d).

IMPACT OF THE OPEN ACCESS NATURE OF MEDLINE® AND MAUDE® UPON THE DEVELOPMENT OF A TWO-SIDED MARKET

Technology platforms such as open access, searchable databases that provide data about patient safety events have identified the presence of technology-induced errors. These platforms have allowed researchers to study and disseminate information about how technology-induced errors can be reduced and mitigated. The dissemination process has been effective. Typically, for information to be used in practice by health professionals (including HIT professionals and health informatics professionals) it takes over 17 years (Morris et al., 2011). In the case of technology-induced errors we saw the first reports in 2005 about this HIT safety phenomenon (e.g., Koppel et al., 2005; Kushniruk et al., 2005). The development, publication, and indexing of articles outlining the strategies needed to address these phenomena began to emerge in 2008 with policy makers identifying the need to encourage reporting of such errors and to address them through the use of regulation in 2011 with the publication of the Institute of Medicine's report on HIT safety (IOM, 2011).

More widespread use of strategies to prevent future occurrences of technology-induced errors is now coming to the forefront. Today, there exist several government websites that have identified regulatory solutions, testing approaches, and methods for disseminating information about how to improve the safety of HIT and to address unsafe features and functions of HIT for healthcare consumers and for health professionals (see Kushniruk et al., 2005). In addition to these, organizations such as the International Medical Informatics Association (IMIA) have published several documents outlining best practices from around the world (e.g., Borycki et al., 2012a,b; Kushniruk et al., 2012).

FUTURE DIRECTIONS FOR A TWO-SIDED MARKET INVOLVING HIT SAFETY

A two-sided market has clearly emerged in the area of HIT Safety. HIT, initially marketed as a tool that could be used to reduce medical error, has now become the focus of attention of healthcare consumers, health professionals, governments, and regulatory bodies. There has emerged a ground swell of interest in vendors and healthcare organizations providing technology that improves patient safety while at the same time having its safety improved (Borycki & Kushniruk, 2008; IOM, 2011; Institute of Medicine 2016). As many HIT are currently being used in healthcare settings, and the safety of these technologies is improving, there continues to be a need for greater access to information through open access searchable databases such as Medline® to disseminate information about technology-induced errors and approaches to mitigating them. There is also a demand among healthcare consumers and health professionals for reporting about technology-induced errors. It is expected that the number of HIT related technology-induced errors will increase as healthcare continues to modernize. Open access platforms such as Medline® and MAUDE® are key to continuing to develop a two-sided market.

Medline® and MAUDE® are platforms that can be used by healthcare consumers, health professionals, and health information technology professionals to provide information about technology-induced errors they have experienced (i.e., actual errors) and aware of (i.e., experienced near misses). Analysis of open access publications and reports on technology-induced errors is critical for researchers and technology developers to understand those features and functions of HIT that lead to patient harm or death and to improve HIT design, development, programming, customization, and implementation.

Such information allows for the development of new risk management and technology-induced error approaches. There is a need for more sharing of information about technology-induced errors between the healthcare consumer/health professional and healthcare organizations who are purchasing and customizing HIT and the vendors who are responsible for the design, development, and implementation of HIT. Full open access data that allows for consumer and health professionals to disclose incident reports involving HIT has benefits that are industry-wide, as specific types of technology-induced errors are common across products and specific features and functions of HIT that are error prone could be improved upon. Such public disclosure of incident reports would support collaborative activity across the HIT industry to improve safety.

HIT is ubiquitous. Today, HIT vendors can be found across continents. Research findings suggest contextual influences are present where technology-induced errors are concerned (i.e., how healthcare is delivered in a specific country will affect the presence or absence of specific types of technology-induced errors). There is a need for platforms for incident reporting to exist at an international or regional level so countries with comparable healthcare systems can learn from each other about the relationship between healthcare work processes, HIT, and health system approaches to providing patient care. Future development of platforms to allow for reporting, analysis, and dissemination of methods to mitigate technology-induced errors will be critical to improving HIT internationally. There are opportunities for countries to learn from each other in the public reporting about technology-induced error incidents and the dissemination of information about mitigating this international problem. Future international efforts should include aggregating technology-induced error data across countries and globally. In addition to this, information needs to be disseminated about technology-induced error solutions—a platform that would allow for the exchange of such information would lead to greater HIT safety over the long term.

In summary, two platforms, Medline® and MAUDE®, have led to the development of a two-sided market around HIT safety. Open access to research and incident reports allowed for health professionals and consumers to learn about technology-induced errors in a few short years, leading to international efforts to report on and create mitigation strategies for addressing technology-induced errors. Over time, it is expected that these platforms will reduce costs associated with technology-induced errors by eliminating the healthcare costs associated with treating patients who have been harmed by a technology, and by healthcare organizations that are implementing these technologies and providing vendors with feedback about the specific types of technology features and functions that are prone to technology-induced errors so they can be addressed during HIT design and customization prior to HIT implementation. Vendors could use such information in their future HIT designs. Healthcare organizations who are customizing and implementing HIT can be informed as to the promises and pitfalls of specific types of HIT used in differing healthcare contexts (e.g., hospitals, physician offices). This would allow HIT vendors and those healthcare organizations that employ HIT to eliminate technology-induced error features and functions prior to implementation and improve the safety of HIT.

Technology-induced errors are a significant healthcare organizational concern as there are significant costs associated with re-implementing systems. For example, health professionals have identified the presence of technology-induced errors in the EHR they are using. In some cases this has led to health professionals petitioning healthcare organizations to investigate the causes of these errors and petitioning both the healthcare organization and vendor to re-design and re-implement the HIT. Healthcare organizations and governments have responded by stating they will fix these systems.

Technology-induced errors are a concern for vendors. Increasingly, there has emerged a trend toward software vendors being sued for technology-induced errors that have led to (as documented in the published literature by Magrabi and colleagues, 2010, 2012) patient harm or death. With the recent modernization of healthcare systems around the world, the number of reports of these types of lawsuits have increased significantly in some countries (e.g., the United States).

There has emerged an incentive for vendors to build safer HIT to reduce the costly expenses of associated lawsuits and settlements, and the re-design, re-programming, and re-implementation after an event involving patient harm. Detailed information about technology-induced errors and near misses (where a technology-induced error may have occurred, but did not as a patient or health professional prevented it) need to be provided. Such information can be used to provide feedback and improve the safety and quality of HIT over time. Technology-induced errors represent an opportunity for organizations (e.g., vendors) wishing to improve the quality and safety of HIT used by healthcare consumers and health professionals as well as private corporations and governments for whom they work. From a HIT industry perspective, instances where technology may lead to a technology-induced error can be observed to be opportunities to improve the safety of HIT over time as each report is studied and the contributing factors are determined and documented and risk mitigation strategies and solutions are developed. Such work can only occur if appropriate platforms are developed for the sharing of such information across the HIT industry.

REFERENCES

Ashish, J.K., DesRoches, C.M., Campbell, E.G., Donelan, K., Rao, S.R., Ferris, T.G., et al., 2009. Use of electronic health records in US hospitals. N. Engl. J. Med. 360 (16), 1628–1638.

Australian Practice Guidelines for Quality and Safety, 2012. Electronic medication management. <http://www.safetyandquality.gov.au/our-work/medication-safety/electronic-medication-management-systems/>.

Baker, P., 2011. Mobile health apps, Part 4: Life, death and lawsuits. TechNewsWorld. <http://www.technewsworld.com/story/72394.html>.

Baylis, T.B., Kushniruk, A.W., Borycki, E.M., 2011. Low-cost rapid usability testing for health information systems: is it worth the effort? Stud. Health Technol. Inform. 180, 363–367.

Borycki, E., Keay, E., 2010. Methods to assess the safety of health information systems. Healthc. Q. 13, 47–52.

Borycki, E., Kushniruk, A., 2005. Identifying and preventing technology-induced error using simulations: application of usability engineering techniques. Healthc. Q. 8(Sp).

Borycki, E.M., Kushniruk, A.W., 2008. Where do technology-induced errors come from? Towards a model for conceptualizing and diagnosing errors caused by technology. Hum. Soc. Organ. Aspects Health Inf. Syst., 148–165.

Borycki, E.M., Kushniruk, A., Keay, E., Nicoll, J., Anderson, J., Anderson, M., 2008. Toward an integrated simulation approach for predicting and preventing technology-induced errors in healthcare: implications for healthcare decision-makers. Healthc. Q. (Toronto, Ont.) 12, 90–96.

Borycki, E.M., Kushniruk, A., Keay, E., Nicoll, J., Anderson, J., Anderson, M., 2009. Toward an integrated simulation approach for predicting and preventing technology-induced errors in healthcare: implications for healthcare decision-makers. Healthc. Q. 12, 90–96.

Borycki, E.M., Househ, M.S., Kushniruk, A.W., Nohr, C., Takeda, H., 2011. Empowering patients: making health information and systems safer for patients and the public. Yearb. Med. Inform. 7 (1), 56–64.

Borycki, E.M., Kushniruk, A.W., Bellwood, P., Brender, J., 2012a. Technology-induced errors. Methods Inf. Med. 51 (2), 95–103.

Borycki, E.M., Kushniruk, A., Nohr, C., Takeda, H., Kuwata, S., Carvalho, C., et al., 2012b. Usability methods for ensuring health information technology safety: evidence-based approaches. Yearb. Med. Inform. 8, 20–27.

Boulos, M.N., Brewer, A.C., Karimkhani, C., Buller, D.B., Dellavalle, R.P., 2014. Mobile medical and health apps: state of the art, concerns, regulatory control and certification. Online J. Health Inform. 5 (3), 229.

Horsky, J., Kuperman, G.J., Patel, V.L., 2005. Comprehensive analysis of a medication dosing error related to CPOW. J. Am. Med. Inform. Assoc. 12 (4), 377–382.

Househ, M., Borycki, E., Kushniruk, A.W., Alofaysan, S., 2012. mHealth: a passing fad or here to stay. In: Rodrigues, J.J. (Ed.), Telemedicine and E-health services, policies and applications: advancements and developments. IGI Global, Hershey, pp. 151–173.

Institute of Medicine, 2011. Health IT and Patient Safety: Building Safety Systems for Better Care. The National Academies of Science, Engineering and Medicine, Washington.

Institute of Medicine, 2016. Diagnostic Error in Healthcare. The National Academies of Science, Engineering and Medicine, Washington.

Koppel, R., Metlay, J.P., Cohen, A., Abaluck, B., Localio, A.R., Kimmel, S.E., et al., 2005. Role of computerized physician order entry systems in facilitating medication errors. J. Am. Med. Assoc. 293 (10), 1197–1203.

Kushniruk, A., Beuscart-Zéphir, M.C., Grzes, A., Borycki, E., Watbled, L., Kannry, J., 2010. Increasing the safety of healthcare information systems through improved procurement: toward a framework for selection of safe healthcare systems. Healthc. Q.

Kushniruk, A., Nohr, C., Jensen, S., Borycki, E., 2012. From usability testing to clinical simulations: Bringing context into the design and evaluation of usable and safe health information technologies. Yearb. Med. Inform. 8, 78–85.

Kushniruk, A.W., Triola, M.M., Borycki, E.M., Stein, B., Kannry, J.L., 2005. Technology induced error and usability: the relationship between usability problems and prescription errors when using a handheld application. Int. J. Med. Inform. 74 (7), 519–526.

Kushniruk, A.W., Surich, J., Borycki, E.M., 2012. Detecting and classifying technology-induced error in the transmission of healthcare data. Proceedings of 24th International Conference of the European Federation for Medical Informatics Quality of Life through Quality of Information.

Kushniruk, A.W., Bates, D.W., Bainbridge, M., Househ, M.S., Borycki, E.M., 2013. National efforts to improve health information system safety in Canada, the United States of America and England. Int. J. Med. Inform. 82 (5), e149–e160.

Magrabi, F., Ong, M.S., Runciman, W., Coiera, E., 2010. An analysis of computer-related patient safety incidents to inform the development of a classification. J. Am. Med. Inform. Assoc. 17 (6), 663–670.

Magrabi, F., Ong, M.S., Runciman, W., Coiera, E., 2012. Using FDA reports to inform a classification for health information technology safety problems. J. Am. Med. Inform. Assoc. 19 (1), 45–53.

Magrabi, F., Baker, M., Sinha, I., Ong, M.S., Harrison, S., Kidd, M.R., et al., 2015. Clinical safety of England's national programme for IT: a retrospective analysis of all reported safety events 2005 to 2011. Int. J. Med. Inform. 84 (3), 198–206.

Morris, R.M., 2009. Encyclopedia of Health Services. Sage, Thousand Oaks, California.

Morris, A.S., Wooding, S., Grant, J., 2011. The answer is 17 years, What is the question: understanding time lags in translational research. J. R. Soc. Med. 104 (12), 510–520.

Mullner, R.M., 2009. Adverse events. Encyclopedia of Health Services Research. Sage, Thousand Oaks, California.

Office of the National Coordinator. About ONC <https://www.healthit.gov/newsroom/about-onc>.

Office of the National Coordinator. Policy making, regulation and strategy. <https://www.healthit.gov/policy-researchers-implementers/health-it-and-safety>.

National Health Service and Microsoft (n.d.). Common user interface. <http://systems.hscic.gov.uk/data/cui>.

Palojoki, S., Mäkelä, M., Lehtonen, L., Saranto, K., 2016a. An analysis of electronic health record–related patient safety incidents. Health Inform. J. 1460458216631072.

Palojoki, S., Pajunen, T., Saranto, K., Lehtonen, L., 2016b. Electronic health record-related safety concerns: a cross-sectional survey of electronic health record users. JMIR Med. Inform. 4 (2), e13.

Rothman, J., 2000. What does it cost you to fix a defect? And why should you care? <http://www.jrothman.com/articles/2000/10/what-does-it-cost-you-to-fix-a-defect-and-why-should-you-care/>.

Samaranayake, N.R., Cheung, S.T.D., Chui, W.C.M., Cheung, B.M.Y., 2012. Technology-related medication errors in a tertiary hospital: a 5-year analysis of reported medication incidents. Int. J. Med. Inform 81 (12), 828–833.

Waterman, A.D., Garbutt, J., Hazel, E., Dunagan, W.C., Levinson, W., Fraser, V.J., et al., 2007. The emotional impact of medical errors on practicing physicians in the United States and Canada. Joint Comm. J. Qual. Patient Saf. 33 (8), 467–476.

CHAPTER 3

Usability of Healthcare Information Technology: Barrier to the Exchange of Health Information in the Two-Sided E-Health Market?

A.W. Kushniruk and E.M. Borycki
University of Victoria, Victoria, BC, Canada

INTRODUCTION

Worldwide there has been a proliferation of health information technologies (HITs) with the aim of modernizing healthcare. This has included considerable spending on systems such as electronic health records (EHRs), decision support systems (DSSs), and many other types of healthcare applications. This has also included the development of a range of personal health records (PHRs) which allow patients and citizens to control access to their own health information to varying degrees. However, there are also increasing reports about the poor usability of many of these HITs (Jamoom et al., 2012; Riskin et al., 2014). This has included concerns about the usability of EHRs, PHRs, decision support systems, mobile applications, and other types of HIT applications (Beuscart-Zéphir et al., 2007; Marcilly et al., 2011; Zhang et al., 2003). Vendors have come under criticism for not responding to the end-user (e.g., clinicians or patients) as well as health organization (e.g., hospital or health authority) concerns about poor usability. In addition, there have been increasing calls for more focus on defining and requiring effective usability for EHRs, PHRs, and related systems (Kellermann and Jones, 2013; Lowry et al., 2014; Riskin et al., 2014). Along these lines, in healthcare a range of electronic platforms have appeared for interconnecting health professionals with patients/citizens and allowing for health information exchange in a two-sided market (Byrne et al., 2014; Irizary et al., 2015; Vest, 2012; Vest et al., 2014). However, a number of issues have limited the effectiveness of initiatives involving access by citizens to their own health data, both at a local level but also at the regional and national levels (Greenhalgh et al., 2010). These issues are related to the need for a usable and context-sensitive display of information for the different markets as well as the need for interoperable network structures that allow for the flow of information across different markets and user types (Edwards et al., 2010).

Despite the problems and issues in implementing HIT, a range of methods exist and can be applied that can help to better understand user needs, as well as to identify and rectify usability problems. These approaches can be used both for improving systems being designed and also for deployed implementations of vendor products. The question remains as to why user issues are still a major problem and what can be done to improve the interconnection between patients and healthcare providers. To address this, in this chapter we will present the case for how improved HIT usability would be of value for both clinical as well as patient/citizen end-users of HIT. It may be argued that only when user needs in this two-sided market are better understood, that we may see major improvements regarding the usability of HIT, along with a more interoperable and useful exchange of health information across both sides of the two-sided e-health market. It will be further argued that usability of healthcare IT is likely to become a critical success factor for organizations promoting proliferation of health technologies. In addition, such interactions between patients/citizens and health professionals should be made context sensitive to respond to the demands and information needs of different user groups.

CONNECTING THE TWO-SIDED E-HEALTH MARKET: THE CASE OF PERSONAL HEALTH RECORDS AND PATIENT PORTALS

Over the past two decades there has been a proliferation of personal health records and health portals designed to allow patients and citizens to access at least a portion of their own health record electronically. Portals are electronic web spaces that allow different user groups (physicians and patients) to share information and supply services that can provide the digital infrastructure for a two-sided market (Irizary et al., 2015). In considering patient access to their health records electronically, we can distinguish between the following (Tang et al., 2006; Halamka et al., 2008): (1) Stand-alone personal health records, (2) tethered personal health records, and (3) interconnected/ interoperable patient portals. Stand-alone personal health records allow patients and citizens to store and retrieve their own health information; however, they do not interface or integrate with information about a patient contained in hospital or institutional records. In contrast, tethered PHRs support a two-sided market by allowing patient and citizen access (often read only) to their records, which are also accessible by their health providers (i.e., this type of personal health record may be tethered to an institutional or hospital record). Moving to the most integrated model is the interconnected or interoperable portal, which may contain and connect up the patient or citizen with information from multiple records (e.g., from clinics, hospitals) and these may be deployed at the regional and even national levels. Some of these systems may allow for citizens and patients to not only read information about themselves (from their health organization), but also add, modify, or otherwise control information about themselves to varying degrees.

HUMAN ISSUES AND SOCIO-TECHNICAL ISSUES IN HEALTH INFORMATION EXCHANGE

From the first documented patient access to their own electronic records over the Internet—the PatCIS project of the late 1990s (Cimino et al., 2002)—there has been a massive increase in direct patient access to their own electronic records. For example, since the early 2000s the Veteran's Affairs (VA) in the United States developed a patient interface to their EHR, known as MyHealtheVet. The system is now accessed daily by hundreds of thousands of users who can access information such as their lab values and reports and who can schedule appointments. Other examples include interfaces to commercial electronic medical record systems such as Epic's MyChart, which much like MyHealtheVet allows patient access to key parts of their record and is used by millions of end-users (Halamka et al., 2008). At the larger regional and national levels, examples include Denmark's Sundhed citizen portal that allows citizens to access much of the patient record. It should be noted these systems are also used by health professionals and form a new channel of communication linking the two-sided healthcare market (Halamka et al., 2008).

Despite the success of many personal health record and patient portal projects there have also been a number of documented failures. There is a clear need for improving the interoperability and alignment between patients/citizens and healthcare providers. One notable example is the HealthSpace patient portal in the UK, where despite large investment into creating an infrastructure for communication, the project was terminated. Issues related to lack of functionality desired by patients/citizens as well as lack of usability have been cited as serious problems leading to the project's termination (Greenhalgh et al., 2010). Along these lines, it has been argued that future systems of this type should take into account user health literacy levels and be designed by applying user-centered design approaches. In Australia, a similar situation has emerged where the national personally controlled electronic health record (PCEHR) has undergone criticism for not meeting user needs, for usability problems, and for lack of end-user buy-in. Again lack of usability and a clear business model have been cited as having led to lower than expected uptake of this new technology (Xu et al., 2014). As a result there has been a call to develop and employ new methods for improving both the gathering of user requirements for such systems and portals, as well as improving the usability of these systems for end-users. The remainder of this chapter will discuss both advances in user requirements gathering and usability engineering for improving the two-sided e-health market.

THE USER-TASK-CONTEXT MATRIX: APPLICATION TO ASSESSING AND DOCUMENTING END-USER INFORMATION NEEDS

It is essential that user needs in the two-sided e-health market be understood in the design and development of systems that will be acceptable, usable, and ultimately adopted by different user groups in the two-sided e-health marketplace. Indeed lack of knowledge of user needs, capabilities, and desires has been implicated in the failure of a number of large-scale e-health projects, including the HealthSpace project in the UK (Greenhaulgh et al., 2010).

In response, Kayser and colleagues have come up with a method for including explicit consideration of citizen/patient information needs in the development of new information technologies such as PHRs (Kayser et al., 2015). The approach builds on Kushniruk and Turner's concept (Kushniruk and Turner, 2012) of the user-task-context matrix, which can be used to drive and guide requirements gathering for complex systems and applications that connect up different end-users, belonging to different sides of the two-sided e-health market (namely patients/citizens and health professionals). Developing a user-task-context matrix begins by the design team brainstorming and delineating who the users of the new or envisaged system will be (e.g., doctors, nurses, patients, the patient's family). Then, for each such class of users, the envisaged tasks they will use the system for are also delineated (e.g., looking up information about one's diabetes or other information). The third dimension—the context dimension—is added by specifying how users are expected to apply a system to carry out tasks in different contexts or settings. Fig. 3.1 shows the user-task-matrix applied to gathering requirements for a personal health record.

The user-task-context matrix can be used for reasoning about user-task-context combinations and to drive creation of scenarios for both designing and testing personal health records and portals. For example, the user dimension can be used to define a class of user (e.g., elderly patient) by specifying the attributes of those types of users. The task dimension specifies the type of tasks those users would be expected to carry out using the system (e.g., reading and interpreting medication information). Finally, the context dimension takes into account contextual issues around using the system (e.g., access to information from home using a mobile application). Kayser and colleagues (2015) describe how they have used the matrix concept to develop a process for designing PHRs and portals that explicitly considers the classes of user, their tasks and contexts of use when designing such systems.

LOW-COST RAPID USABILITY ENGINEERING

Usability can be defined as a measure of ease of use of system, for use in a specific context. The dimensions of usability are the following: (1) learnability, (2) effectiveness, (3) efficiency, (4) enjoyability, and (5) safety (Kushniruk and Patel, 2004). Assessing

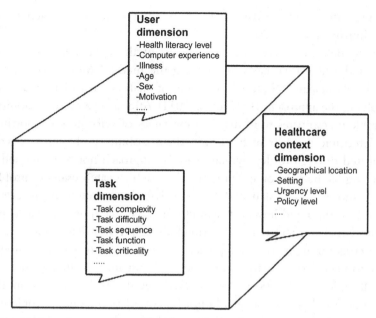

Figure 3.1 User-task-context matrix modified for use in developing PHR use cases.

usability is central to improving systems, including making them more usable and likely to be adopted by different market segments, and usability engineering has emerged as a consequence with scientific methods for measuring usability. Usability evaluation has become central to user-centered design methods, whereby design involves iterative cycles of requirements gathering, prototype development, and evaluation. The authors have been involved in the development and refinement of an approach to usability engineering in healthcare that they have termed rapid low-cost usability engineering (Kushniruk and Borycki, 2006). The approach takes a pragmatic view of usability engineering in HIT and argues that studies can be conducted in situ, at low-cost and rapidly in settings ranging from the clinical setting to home use of technology. The approach involves taking low-cost and portable recording tools into settings where prototypes, developing systems, and completed software products can be tested. This typically involves observing (and video recording) representative end-users of systems (e.g., patients, citizens, and health professionals) as they interact with software tools and systems to carry out typical and representative tasks (e.g., using a PHR to look up their blood glucose trends). This typically involves using free screen recording software to record the complete user interaction with the system/application under study. Participants of such studies are also typically asked to "think aloud" or verbalize their thoughts (which is also audio recorded). The resultant recordings of user interactions can then be played back, transcribed, and coded to identify user problems and come up with recommendations for improving usability for each class of user studied (Kushniruk and Patel, 2004).

In order to determine the cost-effectiveness of the approach, a series of studies were conducted where total costs of applying the method were recorded (e.g., costs of equipment, participant remuneration, costs of analysis) in order to document the costs of applying the approach (Baylis et al., 2011). After analyzing the usability data for usability problems, estimates were made (using tables of software costs postimplementation) to determine the cost of fixing identified usability problems that would have been propagated through to deployment, had the approach not been applied, in order to estimate tangible benefits of applying the approach to catch usability problems. The cost–benefit ratio was then calculated. From analyses of the results from several usability studies where the rapid low-cost usability engineering approach was applied (and the costs and benefits calculated as described above), it was determined that benefits outweighed costs for applying the approach in all cases, with percent savings ranging from 36.5% to over 200%. Furthermore, the total cost for conducting the studies was typically well under US$10,000 per study (Baylis et al., 2011). In addition, the methodology also detected errors that would not be considered usability problems, including programming bugs, safety issues, and technology-induced errors (that would have otherwise not have been detected prior to widespread implementation).

As poor usability has become recognized as being one of the greatest barriers to effective implementation of healthcare IT today, including patient portals, such integrated approaches to detect serious usability problems have begun to be developed that can serve as a safety net for catching problems. However, despite the advent of methods such as those described above, the issue remains as to why there continue to be many reported problems about usability. One argument is that there is the need for greater education and dissemination of usability engineering methods, best practices, and evidence-based health informatics. A second argument is related to the economic value of designing more usable systems, as described above regarding cost–benefit analyses where it was shown that application of currently available methods may have a big impact on end-user satisfaction. These constitute issues and constraints that are limiting the value of PHR and EHR investments to citizens, patients, clinicians, and healthcare organizations, including the need to show the value proposition of designing more usable and effective EHRs from the vendor's perspective as well.

APPLICATION OF LOW-COST RAPID USABILITY ENGINEERING THROUGHOUT THE SYSTEM DEVELOPMENT LIFE CYCLE (SDLC)

Usability should be considered throughout the full life cycle of systems that are designed to support communication and interaction between citizens/patients and health professionals. This is especially true of more complex applications and systems such as EHRs and PHRs designed to provide communication between different sides of the e-health market. The system development life cycle (SDLC) provides a framework where application of usability engineering methods can be considered. This includes methods for gathering improved user requirements in early stages of the SDLC during requirements gathering. The methods can be used in conjunction with the user-task-context matrix idea described above, whereby scenarios for user testing are created from the user-task-context matrix. Such scenarios form use cases for both specifying user requirements for systems (e.g., patient–citizen requirements as well as health professional requirements) and for driving the usability testing during user-centered design.

It should be noted that usability engineering methods can be used during early design of systems to compare potential user interface layouts and designs. Later in design usability engineering methods can be used to test and refine user interface prototypes, in a cycle process of user-centered design involving cycles of design, and evaluation of design ideas using prototypes shown to end-users (which can involve applying low-cost rapid usability engineering methods). Later in the SDLC, low-cost rapid usability methods can be used to refine user interfaces and systems during system implementation and deployment. Finally, the methods can be used to see if the designed systems meet user needs when fully deployed in the marketplace. Along these lines, such testing should be mandatory as it will uncover not only usability problems, but also potential for technology-induced error, such as entry of wrong medication information into a PHR by a patient user (Borycki and Kushniruk, 2005).

Even in the case of procurement of health information systems (where the buyer does not design the system from scratch but rather a vendor system is purchased with limited ability for the buyer to change its structure), the methods described above have been shown to be of considerable value. For example, Kushniruk and colleagues (2010) describe a continuum of evidence for fit between a candidate system (for procurement) and the buying organization (e.g., a hospital or health region) from the end-user's perspective. This approach argues for requiring a test install of candidate systems (e.g., EHRs) and usability analysis of the interactions of users belonging to different market segments. Based on empirical data (i.e., comparing which system best fits organizational and end-user needs) as well as end-user comparison of the candidate products, it has been shown that better procurement decisions can be made (Kushniruk et al, 2010).

DISCUSSION

The work described in this chapter has considered user needs and usability in the context of the two-sided e-health market. With the widespread deployment of systems intended to connect different sides of the e-health market (e.g., patients and citizens with healthcare professionals and health organizations), the issues of user needs and usability have come to the fore. A range of methods can be applied to better describe and assess user needs in the development of new platforms intended to provide patient information across portals and to allow for interoperable access to key patient information. However, there are currently a number of examples of failed efforts internationally which were aimed at such integration of information. Critical examination of these projects has indicated a lack of consideration of user needs and in particular the requirements of users from different sides of a two-sided market. Furthermore, the issue of lack of usability of systems has also come to the fore, leading to the argument for greater emphasis on developing effective and usable systems for all classes of end-users. Toward this goal, advances in usability engineering as applied to HIT promise to lead to improved design of systems to meet user needs, improved uptake of e-health systems and increased benefits of their deployment. The methods are readily applicable for rapid application in the design and evaluation of a range of HITs, including those targeted to both patient and health professional users. In conclusion, human barriers to the exchange of health information will require more intensive and continued consideration of both user needs and usability.

REFERENCES

Baylis, T.B., Kushniruk, A.W., Borycki, E.M., 2011. Low-cost rapid usability testing for health information systems: is it worth the effort? Stud. Health Technol. Inform. 180, 363–367.

Beuscart-Zéphir, M.C., Elkin, P., Pelayo, S., Beuscart, R., 2007. The human factors engineering approach to biomedical informatics projects: state of the art, results, benefits and challenges. IMIA Yearb. Med. Inform., 109–127.

Borycki, E., Kushniruk, A.W., 2005. Identifying and preventing technology-induced error using simulations: application of usability engineering techniques. Healthc. Q. 8, 99–105.

Byrne, C.M., Mercincavage, L.M., Bouhaddou, O., Bennett, J.R., Pan, E.C., Botts, N.E., et al., 2014. The department of veteran's affairs (VA) implementation of the Virtual Lifetime Electronic Record (VLER): findings and lessons learned from Health Information Exchange at 12 sites. Int. J. Med. Inform. 83 (8), 537–547.

Cimino, J., Patel, V., Kushniruk, A.W., 2002. The patient clinical information system (PatCIS): technical solutions for and experiences with giving patients access to their electronic medical records. Int. J. Med. Inform. 68 (1-3), 113–127. 18.

Edwards, A., Hollin, I., Barry, J., Kachnowski, S., 2010. Barriers to cross-institutional health information exchange: a literature review. J. Healthc. Inf. Manage. 24 (3), 22–34.

Greenhalgh, T., Hinder, S., Stramer, K., Bratan, T., Russell, J., 2010. Adoption, non-adoption and abandonment of a personal electronic health record: case study of HealthSpace. Br. Med. J. 341, c5814.

Halamka, J., Mandl, K.D., Tang, 2008. Early experiences with personal health records. J. Am. Med. Inform. Assoc. 15 (1), 1–7.

Irizary, T., Dabbs, A., Curran, C., 2015. Patient portals and patient engagement: a state of the science review. J. Med. Int. Res. 17 (6), e148.

Jamoom, E., Beatty, P., Bercovitz, A., et al., 2012. Physician adoption of electronic health record systems: United States, 2011. NCHS data brief, no. 98. National Center for Health Statistics, Hyattsville, MD.

Kayser, L., Kushniruk, A., Osborne, R., Norgaard, O., Turner, P., 2015. Enhancing the effectiveness of consumer-focused health information technology systems through e-health literacy: a framework for understanding user needs. J. Med. Internet Res. Hum. Factors 2 (1), e9.

Kellermann, A.L., Jones, S.S., 2013. What will it take to achieve the as-yet-unfulfilled promises of health information technology. Health Affairs 32, 63–68.

Kushniruk, A., Borycki, E., 2006. Low-cost rapid usability engineering: designing and customizing usable healthcare information systems. Healthc. Q. 9 (4), 98–100. 102.

Kushniruk, A., Turner, P., 2012. A framework for user involvement and context in the design and development of safe e-Health systems. Stud. Health Technol. Inform. 180, 353–357.

Kushniruk, A., Beuscart-Zephir, M.C., Grzes, A., Borycki, E., Watbled, L., Kannry, J., 2010. Increasing the safety of healthcare information systems through improved procurement: toward a framework for selection of safe healthcare systems. Healthc. Q. 13, 53–58.

Kushniruk, A.W., Patel, V.L., 2004. Cognitive and usability engineering approaches to the evaluation of clinical information systems. J. Biomed. Inform. 37 (1), 56–76.

Lowry, S., Ramaiah, M., Patterson, E., et al., 2014. Integrating electronic health records into clinical workflow: an application of human factors modeling methods to ambulatory care. Natl. Inst. Stand. Technol.

Marcilly, R., Leroy, N., Luyckx, M., Pelayo, S., Riccioli, C., Beuscart-Zéphir, M.C., 2011. Medication related computerized decision support system (CDSS): make it a clinicians' partner. Stud. Health Technol. Inform. 166, 84–94.

Riskin, L., Koppel, R., Riskin, D., 2014. Re-examining health IT policy: what will it take to derive value from our investment? J. Am. Med. Inform. Assoc. 0, 1–4.

Tang, P.C., Ash, J.S., Bates, J., Overhage, J., Sands, D., 2006. Personal health records: definitions, benefits, and strategies for overcoming barriers to adoption. J. Am. Med. Inform. Assoc. 13 (2), 121–126.

Vest, J.R., 2012. Health information exchange: national and international approaches. Adv. Health Care Manage. 12, 3–24.

Vest, J.R., Kern, L.M., Campion Jr., T.R., Silver, M.D., Kaushal, R., 2014. Association between use of a health information exchange system and hospital admissions. Appl. Clin. Inform. 5, 219–231.

Xu, J., Gao, X., Sorar, G., Croll, P., 2014. Current status, challenges, and outlook of e-health record systems in Australia Knowledge Engineering and Management. Springer, Berlin, Heidelberg. pp. 683–692.

Zhang, J., Johnson, T.R., Patel, V.L., Paige, D.L., Kubose, T., 2003. Using usability heuristics to evaluate patient safety of medical devices. J. Biomed. Inform. 36 (1), 23–30.

CHAPTER 4

Inclusive Design in Ecosystems

J. Mitchell and J. Treviranus
OCAD University, Toronto, ON, Canada

E-HEALTH, INCLUSION, AND DESIGN

The most recent round of technological disruption has transformed the way we work, communicate, participate, learn, teach, get and share information, think, and view ourselves as human beings. In many cases digital and networked technologies have transformed our interactions with each other, have helped us achieve greater efficiency, have democratized information, and have enabled more equitable participation. Those impacts are often referenced when we talk about the power of technology to lead to a more equitable world. With the ubiquity, speed, and maturity of new technology and the possibilities of deep connectedness, we seem to have important elements for addressing fairness. New technologies have created the opportunity to help individuals leap forward who have historically lagged behind. However, there are many for whom these technological promises are still not realized—there are limits within our increasingly digitized world.

There are some who have no access because of limits in infrastructure caused by geography, economics, or policy. Even if Internet access were truly ubiquitous and affordable across the globe, many would be excluded by the design of interfaces, devices, and content. These individuals face an ever-widening technology gap that is complicated and reinforced by the economics of design, which favor the largest customer base. This accelerated rate of change and preference for the majority is making it easier to be left behind, risking widening the gap. Within that widening gap are those excluded from the digital world of meaningfully doing, thinking, sharing, and being in this new context—the digitally excluded.

When digital exclusion occurs in the healthcare sector, it can be catastrophic. Digitally marginalized individuals suffer greater numbers of negative health outcomes and comorbidities. Something as simple as an e-health portal meant to empower patients to self-manage care, while empowering to some, can prevent access simply by not meeting the unique needs of others. If the service has been designed in a way that does not consider the spectrum of unique needs, it can leave some unable to manage their own health. They are robbed of access but also of participation in an immensely important and intensely personal activity.

E-Health Two-Sided Markets.

This is particularly troubling because those who are digitally excluded are often persons with disabilities and are disproportionately individuals with significant social and economic barriers (low income, lack of access to infrastructure, literacy, etc.) (Thomas, 2012). Digital exclusion to healthcare services perpetuates cycles of exclusion and marginalization that these individuals already experience. As Khaled Abdel Kader, an assistant professor of medicine at Vanderbilt University, states, "Despite the increasing availability of smartphones and other technologies to access the Internet, the adoption of e-health technologies does not appear to be equitable. As we feel we are advancing, we may actually perversely be reinforcing disparities that we had been making progress on" (Shallcross, 2015).

The opportunities for the inclusive design of digital interfaces and complex systems

This chapter focuses on two areas where users are excluded: the first is in the design of digital interfaces (applications, interfaces, websites, kiosks, etc.), and the second is in the complex systems that are aggregations of these limited (and often inaccessible) interfaces.

As the world continues to go digital, there needs to be a strategy or approach to address exclusion and the widening technology design gap. An inclusive design approach can help us address the gaps. By adopting an inclusive design perspective, we can build interfaces that meet individual needs and build the infrastructure for systems that evolve and grow, meeting diverse contexts and requirements. We can build applications that can be used by anyone, not just those with certain abilities who are in particular contexts. And in doing this we can continue to break down barriers to participation. As Eve Andersson, head of Google Accessibility, states: "inclusive design means more than just hacking an app or product so that people with disabilities can use it. It's something that benefits literally everyone" (Brownlee, 2016). Inclusively designed solutions result in better solutions.

In sections two and three we will discuss how the perspective change that comes from following the principles of inclusive design can break down barriers in the field of healthcare—the "next frontier" for modernization and improvement. This shift can have a profound impact on healthcare. As articulated in the US Government National Council on Disability report on the Current State of Healthcare for People with Disabilities, "key stakeholders from diverse communities highly recommended that principles of universal[1] design be applied in all aspects and venues of healthcare, ranging from facilities design and construction to the development of quality measures, research design, and delivery of care that embrace everyone, including people with disabilities" (Vaughn, 2009).

Addressing design limitations in interfaces and the complex systems they make up will address many immediate causes of digital exclusion in e-health. Furthermore, if those complex systems are made more inclusive it creates a foundation upon which an innovative new model can be built. With an inclusive foundation the healthcare sector can move toward a multisided platform or market—one that empowers, involves, and engages individuals in the "design" of their own care and outcomes.

The opportunities for the inclusive design of two- or multisided markets

Traditional markets privilege the average consumer or largest customer base. In traditional markets services, development, and innovation are driven by big data, majority needs, and addressed with a one-size-fits-all approach (Treviranus, 2014). This causes harm to both innovation and consumers. Traditional markets leave consumers and producers in a stagnant state where information is fragmented, there are inefficiencies and redundancies, and there is a lack of discoverability of solutions that meet individual needs. In healthcare that means individual care is addressed from an interpretation of the typical or average presentation, diagnosis, and medication administration (evidence-based medicine). This is correlated with poor outcomes, unmet medical needs, health disparities, or dangerous risks for patients not conforming to the norm.

The healthcare field acknowledges the inherent limitations of the traditional approach (Epstein and Street, 2011 and Crossing the Quality Chasm, 2001). This can be seen in the popularity of personalized medical approaches (e.g., patient-centered medical home, patient- and family-centered care, pharmacogenomics, and more) that deal with individuals as agents in specific family, social, and economic contexts. This shift requires and supports a more inclusive approach to healthcare, one where the individual is unique and diverse, must be understood individually, and must be an active participant in his/her care. It is an approach where the tools used in care support individual uniqueness and need, and empower the individual to be an agent of their own outcomes. It is an approach that creates the conditions for innovation and experimentation.

Section four will examine how an inclusive ecosystem made up of inclusive interfaces and systems can support a multisided platform for healthcare. We will explore how a multisided market can help address the extant health disparities, unmet medical needs, and poor outcomes by addressing exclusion on all levels—ultimately ending cycles of exclusion while broadening market possibilities.

AN INCLUSIVE DESIGN APPROACH

Inclusive design is design that considers the full range of human diversity with respect to ability, language, culture, gender, age, and other forms of human difference (Treviranus, n.d.). It is good design that can be understood through three main dimensions:

1. Recognition and respect for human diversity, self-determination, and the uniqueness of each person
2. Use of inclusive processes and tools to support and leverage the participation of diverse perspectives and enable participation by the intended beneficiaries of the design
3. Cognizance of the systemic impact of design and work toward positive systemic change

Each of these three dimensions will be examined and will be threaded throughout this chapter as we first explore the inclusive design of interfaces and systems and then explore the opportunities of multisided markets in health (Treviranus, 2016).

Recognize human diversity and uniqueness

Each of us is unique in and of ourselves, and changing from moment to moment. The context around us is also ever-changing, adding additional complexity to understanding our behavior, preferences, and needs. While many of our technologies attempt to "know" what we want and need or how we feel, ultimately only we ourselves can fully know and potentially express these things. Technology that tries to anticipate what we need, think, and feel can only offer a best guess based on sophisticated but limited algorithms that attempt to reduce our complex selves to probability and pattern and may well rob us of potential self-knowledge.

Inclusive design is design that aims to recognize the uniqueness and diversity in all of us, and wherever possible let each individual make decisions for themself. By extending agency to individuals, inclusive design puts an end to cycles of exclusion and technology that strip us of choices. The burden is shifted from an individual to behave predictably and typically to an interface to be designed with diversity, complexity, and the unexpected in mind. The practical implications of this for designing systems and interfaces are that users must be able to make choices and declare preferences wherever possible. One way to make that possible is to prioritize personalization. A one-size-fits-one approach empowers users to declare what works uniquely for them in their changing context—in other words, let the user decide wherever possible. This can translate into practicable steps in the design process where designers can acknowledge human diversity by offering flexibility and customizability or personalization whenever possible.

What is personalization though? It is intended to be different for everyone, so designs must be flexible and make as few assumptions as possible. Design decisions risk being exclusionary if they aren't made with diversity in mind. If I as a designer of an interface put a blue button in the upper left of my page to serve a particular function, I've made potentially exclusionary decisions with each declaration: blue and not any other color, upper left and nowhere else, button and no other affordance, and only this function. Those four decisions:

- Color = blue
- Presentation = button
- Location = upper left
- Function = defined function
  can each exclude potential users from the interface.

To design with a one-size-fits-one perspective is to question each of these granular design decisions from a perspective of inclusion and human diversity. For example, can I imagine how any of the decisions would make the interface unusable to someone? Can users with low vision or color blindness see the blue I have chosen? Is the button usable by someone with alternative input devices like the keyboard or a touch screen (does it just work for a mouse, is it big enough to make selection on a touch screen straightforward)? Is a button the appropriate affordance to use (e.g., does a toggle or a slider make more sense)? From a design perspective, is it intuitive that this particular feature is in the upper left? Will that meet the needs of someone who is distracted or someone who requires a simplified version? And does the defined function meet the needs and expectations of those who require it? By asking these questions, the designer can build an interface that possesses a degree of inherent flexibility to meet many diverse needs. Without this perspective, users can easily be excluded from access because of something seemingly small.

Mismatch

Inclusion is about human difference and one way often used to describe differences among people is by articulating their differences in terms of ability or disability. We often hear about disability within the context of a medical condition—someone is afflicted and cannot do something because of a permanent, temporary, born-with, or acquired deficit. These deficits are sorted into four categories that make up the medical model of disability: cognitively impaired, hearing impaired, sight impaired, mobility impaired.

An inclusive design approach redefines the concept of disability. Disability is a mismatch between the needs or preferences of a user in a particular context and a design (whether it is a service, product, process, interface, technology, or environment) they are interacting with (Ayotte et al., 2014). Mismatch can lead to an inability to access information, services, or technology, and it can also manifest as a discomfort, difficulty, or suboptimal experience with an interface or service. With this change in perspective we can see it is the interface that is failing, not the user. Accessibility of a user interface or service is measured by its ability to meet the needs and preferences of diverse individuals in diverse contexts.

To illustrate *mismatch* consider the story of Emerald (Burns 2015), a 37-year-old mother of six about to give birth to her seventh child. Emerald's family lives in Judith Gap, Montana, a rural farm community that is two hours' drive away from the nearest medical clinic. The distance has limited Emerald's access to prenatal care, and it also presents a significant problem for her baby's birth. Emerald had an emergency cesarean section with her sixth child because the cord had prolapsed. She was lucky that someone earlier that night had come into the clinic with a broken leg because otherwise there would not have been an anesthesiologist available for her surgery and her baby would likely have died. All five of Emerald's other births were vaginal and she found the cesarean very painful and difficult to recover from. And it was important that Emerald get back on her feet quickly to take care of her family. With her seventh birth she wants to choose a vaginal birth (known as a VBAC, vaginal birth after cesarean).

Insurers will often not cover VBACs because it is considered higher risk and if something were to happen during the birth it puts the physician at a higher risk of a costly malpractice suit.

Emerald's decision about her own care is effected by a number of circumstances that are unique to her context (she does not live close to a clinic), she has a number of children (she needs to get back to her family quickly after the birth), and she prefers vaginal births (her birth is considered higher risk). Other contextual factors that effect Emerald are that she lives in a country with over double the usual rate of cesarean births (according to the WHO 2015b). Emerald also lives in a country where insurers often will not cover procedures that they deem potentially costly. In many ways these contextual details result in a mismatch between how Emerald would choose and what is made available to her. While some of Emerald's mismatches are entangled in complex systems, some are quite solvable.

Design-addressable

The redefinition of disability and accessibility leads to an important perspective shift. The reframing of accessibility puts the burden (and opportunity) on design—access and accessibility become tractable design and development challenges that impact us all. Since disability may be a momentary or contextual mismatch, all design work must thus consider a range of potential mismatches, and must take into account a diversity of needs and preferences. Objections could be raised as to the feasibility of this approach given the huge diversity and complexity of requirements. One temptation might be to address some diverse needs with a separate solution. The perspective shift is not practicable if the diverse needs are addressed in a segregated fashion. Segregated approaches are not sustainable, inevitably leaving minority approaches orphaned or underserved and again perpetuating cycles of exclusion. The system as a whole must be designed to address diversity. In doing this, the designs become more usable and accessible for everyone.

The best way to accomplish this kind of design is to engage users early in the process to gain an understanding of their unique and changing requirements. This shift in perspective engages the end-user in understanding and crafting a design that is "right" for them. Inclusively designed interfaces and services allow the user to personalize their own experience—the user continues and contributes to the design, making adjustments to meet their own unique needs and preferences. This shift and the insertion of personalization provide the design foundations to make a system that is highly usable, adaptable to diverse user needs and goals, and that allows for growth and innovation prompted by unique uses.

Processes and tools

The second dimension of Inclusive Design is the use of inclusive processes and tools. By using tools that encourage and allow anyone to participate and be represented, the first dimension of inclusive design can be accomplished—more diverse perspectives can be represented and involved in the design process when decisions are made.

Within the disability community this codesign process is captured with the saying "nothing about us without us" (Charlton, 2000). In other words, don't guess at what the users want or need or prefer. Let the users decide and involve users in the early stages of (and throughout) the design process where these details help to form the final interface design.

Some questions to ask to achieve the second dimension of inclusive design are:

- Who can participate?
- Who is excluded?
- Are these individuals excluded because of a limitation of the tools and processes?
- Can those tools be changed or are there tools that can be used to allow everyone to participate?

Inclusive design processes make use of a number of inclusive practices, principles, tools, and activities.[2] There is no fixed list of tools, nor a fixed process to follow, rather a commitment to inclusion. Inclusive design is iterative, adapting, and growing so the ways to do it and the tools to use will change to reflect evolving thinking, to adapt to the context for the design challenge, and to reflect the approach of the practitioner.

Codesign

When you shift your design perspective to recognize human diversity and uniqueness, those diverse perspectives must be part of the design process. Great design ideas can come from anywhere and especially from individuals' expression of their own uniqueness, so it is essential that end-users are part of the design activities from the beginning. Great design is not something only experts with design degrees can conduct. Rather, each of us has experience with interfaces that work, and those that do not.

Inclusive design evolves the role of the "user" from being observed and used as inspiration to being a designer from the beginning. Personalized solutions can best be designed and iterated with the active participation of the individuals whose needs they are meant to address. In this approach, the user is a codesigner in the design process.

The role of the designer shifts from being tasked with brilliant design decisions to acting as a facilitator and interpreter of design ideas—a translator with a specialized set of tools and training. The designer possesses a wealth of experience and knowledge in interfaces, individual uniqueness, and mismatch. The role of the designer is to remain ever more vigilant that the designs will resolve user mismatch, so they should think as broadly as possible throughout the design process.

For the designer, "the potential for making existing health disparities even worse should be 'panic inducing,' … given the high burden of chronic disease in vulnerable populations. But that presents an untapped opportunity as well: to design products and services tailored to those people" (Shallcross, 2015). Codesign is one way to harness this opportunity. Now "modern services are too frequently designed by educated professionals and policy makers for 'People Like Us', and thus still fail to serve the disadvantaged and societally disconnected, even though they are known to have greater health needs" (Rigby et al., 2015). "The near-pervasive introduction of e-health systems, and the more recent implementation of systems intended for patient use offer patients the opportunity to participate in their own care. Unfortunately the design of these systems means that they may work better for 'People Like Us' rather than for those on the wrong side of the 'digital divide'" (Showell and Turner, 2013). The very individuals that most need interventions to reach them and address the gaps they experience are often excluded from these innovations.

There are efforts to bring individuals and families back into health decision-making. Groups like The Institute for Patient- and Family-Centered Care focus on "promoting collaborative, empowering relationships among patients, families, and healthcare professionals, IPFCC facilitates patient- and family-centered change in all settings where individuals and families receive care and support" (IPFCC, n.d.). This group works to involve patients and families in all aspects of healthcare and health management.

User modeling tools

User modeling tools can be helpful in addressing the first dimension of inclusive design by keeping in mind a range of diverse users. Common tools for modeling users are personas and scenarios. An inclusive design approach to creating personas and scenarios involves describing individuals that have "edge" requirements or needs and contexts at the margins, rather than typical or average scenarios.[3] These personas or scenarios may stand in as proxies if codesigners with these edge needs and contexts are not available. This helps in understanding the scope, uses, and success criteria of the design challenge being addressed. This is often conducted at the beginning of a design process, the point at which the articulation of a diverse range of users can help ensure that the thinking about use is broad and diverse with respect to inclusion and accessibility. These tools can also help designers anticipate weaknesses in the architecting and development of the design—if it cannot meet the needs of diverse users, then designers can address the limitations early on.

Inclusive design mapping tool

The Inclusive Design Research Center has created new design tools to further communicate and explore user uniqueness as inspiration for personalization and flexibility. Exploring uniqueness is essential to an inclusive design approach. Understanding where unique preferences or needs overlap can create unique design opportunities. For example, a busy mom or executive can benefit from the same simplification of content as someone with a learning disability or a cognitive impairment. "The Inclusive Design Mapping tool" (Fluid Project Wiki, 2016b) is a tool created by the Inclusive Design Research Center that helps to visualize the needs, preferences, and contexts of

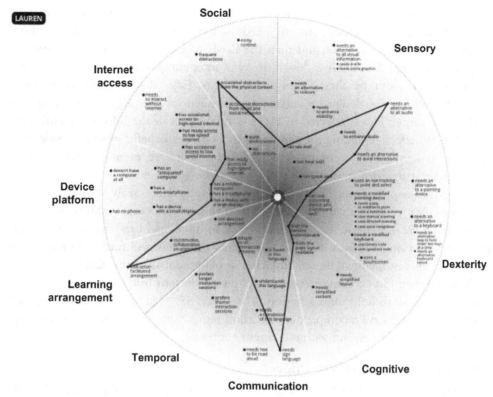

Figure 4.1 Inclusive Design Mapping Tool showing a user constellation.

users over time, and the ability of interfaces to meet those needs. It is a modeling tool for visualizing the particular needs and preferences that the candidate design encompasses. It can be used to map a current product, consecutive iterations of candidate designs, and the varied requirements of different intended users. The goal is to stretch the design to encompass as broad a range of needs as possible (Fig. 4.1).

For the purposes of the tool, *state* is defined as any personal factor that determines the functional requirements that enable optimal use (e.g., sensory, dexterity, cognitive, communication, and temporal), and *context* is something external to the individual that also determines the functional requirements that enable optimal use (e.g., learning arrangement, device platform, internet access, physical environment, and social situations). Each individual represents a jagged, evolving spectrum of states and contexts.

> Because a user's state and context shifts throughout the day, week, and life of a user, what may be true at one moment may not be true in another. For instance, a user's needs for a given product in the morning at the office may be different in the afternoon in the car. This tool attempts to capture the space of these varying states and contexts in order to make the considerations for designing all needs more transparent" (Fluid Project Wiki, 2016a)

The mapping tool places the "typical," "average" needs (or needs that are met by current designs) in the middle, and the needs of individuals that have difficulty using the current design further out, and then the needs of individuals that cannot use the designs at the edge. An inclusive design stretches to encompass the functional requirements brought about by these states and contexts. Edge designs tend to be more diverse. Stretching the design to encompass the needs at the edge encompasses the needs at the center. This is referred to as the transfer effect or the "electronic curb-cut" (Jacobs, 2016) effect: designing for people with disabilities benefits everyone. Sidewalks that were redesigned to accommodate wheelchair users are useful to those with a shopping cart, baby stroller, wagon, bicycle, wheeled luggage, etc.

Using the mapping tool to trace the lines of many users easily shows overlap in needs regardless of individual differences. This helps reveal commonalities and differences among users. The tool can also show how an individual user's needs evolve over time within varying contexts. The insights gained from this tool further reinforce the perspective shift of inclusive design, inspiring development solutions that meet diverse needs rather than just "common" needs. With visual outputs from user states and contexts, it is difficult to hold onto a notion of an "average" need since even an imagined "average" user will have a complex constellation representing their movement through time, to different devices, and within different contexts.

Having a broader beneficial impact

The third dimension of inclusive design is to maintain a sense of a broader beneficial impact. This dimension encourages everyone to acknowledge the larger context. In Emerald's case described above there were many complex factors at play: geography, economic forces, social forces, etc. The larger context is often made up of entangled and interacting "verticals" that make up our world like technological, economic, social, cultural, market-based, and political systems.

The larger context can also be described from a goal-based perspective as has been done in the EU with Disability Strategy Goals: accessibility, equality, education and training, health, external action, employment, social protection, and participation (European Disability Strategy, 2010). With this larger, global view designers can at least have an awareness and sensitivity to the diverse challenges around the world and can set out to solve for everyone. This way, designers will be informed about differences and be able to know and speak directly to whatever constellation of individuals they feel will be implicitly excluded from the solutions (e.g., if we are building a website but we know that some people in the world do not have access to the Web we can create a site that exports easily to a printable format).

This dimension of inclusive design is a reminder to design solutions that will have a far-reaching, thoughtful impact of virtuous rather than vicious cycles. This dimension also directly addresses the fact that many of our solutions should be considered on a global scale.

Once a designer builds something, it will become part of a larger system. And though the design process can determine specific functionalities for that solution and even imagine a target user group, we cannot predict and control who will use the solution and how it will be used. Designers need to take this into account and be thoughtful about how their solution might function for an unintended target group in an unintended market or context.

Applying the three dimensions of inclusive design to healthcare

First, we are all unique—our healthcare decisions will be contextualized by our families, our caregivers, their ability to participate in our care, our ability to understand our doctors (do we speak the same language?), and our ability to participate in our care (follow instructions, manage medication, etc.). Our context matters here too—our culture, the hospital we are in: Is it a trauma 1 center or a rural care center? Is it the end of the month when new residents come in or is it midmonth? Is it when a shift change is about to happen? Medication management is a point of personalization too, namely, pharmacogenomics. Do I typically have strong reactions to low doses of medicine? The experience of the individual in a care facility is complex as well: is there a transfer happening? How will continuity of care be accomplished in a transfer? Is it an intensive care unit, a rehabilitation facility, a long-term care facility?

The way the environment is shaped is another point of opportunity for inclusive design. To see an interesting (inclusive) perspective on some of the physical limitations of hospital environments, see the work of Michael Graves, a designer who found himself compelled to have an impact on the physical space in a hospital room (TEDMED, 2011). Some additional areas where inclusive design can have an impact are:
— Where things break down (communication, function, intention, etc.)
— Where things can be done better (efficiency, accuracy)
— Where things go wrong (response to problems)
— Where things are confusing and complicated

Each of these areas has multiple points of failure or points of mismatch. In each of them an inclusive design perspective could improve outcomes. Healthcare, health outcomes, and hospitals are ripe candidates for an Inclusive Design makeover.

Second, the tools we have available to us will also depend on the hospital, our insurance, our country, our coverage (are we traveling abroad?), the education of our practitioner, the central decisions of our facility, etc. The field of healthcare has encountered many barriers to adopting electronic medical records (Ajami and Bagheri-Tadi, 2013), and is working to maximize the potential of having flexible, digital data. There is also a burgeoning field of self-acquired data—the quantified self. Many people wear devices like Fitbit or have smartphones that are collecting data about individual health: steps per day, sleep habits, calories, etc. We will see many more systems developed that empower patients to learn about themselves, their context, and their world and to manage their own health armed with that information.

Another area where an inclusive design approach is essential is the preservation of privacy and the protection of data (medical and personal) that is being collected and shared. With open data comes great possibilities and also great risk for misuse. The Global Public Inclusive Infrastructure project (GPII Website, 2011) and the IDRC are currently developing interfaces to allow individuals to declare needs and preferences, store them in the cloud, and easily determine who and what can see them (individuals and devices).

The bigger picture, or third dimension of inclusive design in healthcare, is a recognition that we are engaging with a number of complex systems when we look at health (healthcare, medication management, pharmaceuticals, communication, income disparity, geographic disparity, discrimination, etc.) where a number of important decisions get made within a social, political, and economic context. It is important to keep this perspective in mind when addressing an application, process, interface, or function within the healthcare field.

Inclusive design processes should be used and will have a profound impact on the following innovative areas coming within health:

- Methods, techniques, and technologies for the support of healthcare in an increasingly connected environment (e.g., communities of interest and "The Internet of Everything")
- Software and hardware development for privacy preservation and protection of medical data
- Creating alternatives to visual analytics to enable analysis of multimodal and multisensorial data presentation and analysis
- Personalization of user interfaces for increased user engagement and application usability
- Innovative clinical decision support mechanisms for the support of patient self-management approaches
- Patient education approaches that incorporate new technologies and multimodal approaches to learning
- Computer interfaces such as Virtual Reality and Gamification to promote health knowledge and healthy behavior

COMPLEX SYSTEMS AS A FOUNDATION FOR MULTISIDED MARKETS

The WHO's key facts on disability and health states:

- Over a billion people, about 15% of the world's population, have some form of disability
- Between 110 million and 190 million adults have significant difficulties in functioning
- Rates of disability are increasing due to population ageing and increases in chronic health conditions, among other causes
- People with disabilities have less access to healthcare services and therefore experience unmet healthcare needs (WHO, 2015a)

In healthcare this leaves many people in stressful situations where they are unable to find the help they need in a critical time. And with nowhere to turn to find solutions, their needs go unmet and their outcomes continue to be negative.

Inclusive design can leverage emerging technologies to address this critical and widening disparity. By using an Inclusive Design approach, the unmet needs of individuals caught in the digital gap can be addressed, and as has been suggested, the resulting innovations will benefit everyone.

Impacting prosperity

What does it mean to keep in mind a broader beneficial impact while designing? In part it means acknowledging the complex world as the context that all the unique end-users call home. And in acknowledging this, inclusive design addresses head-on the complexities in our ever-changing world—a world made up of complex systems that require novel approaches to design if we are to have an impact on participation, inclusion, health, and prosperity.

Prosperity4All, a European Commission funded project (Prosperity4All, n.d.), is adopting an inclusive design process to address the needs of a diverse population and at the same time create an infrastructure that creates the conditions for innovation. Prosperity4All will enable a new ecosystem and marketplace to grow—one that encourages collaboration, reduces redundant development, lowers costs, increases market reach internationally, and addresses diverse and currently unmet needs. The goal is for this new infrastructure to create an opportunity for yet unrealized technical innovations that lower the bar for participation in development, promote accessibility as a ubiquitous service, and integrate auto-configured features that lower the barrier to access and use of mainstream products. To accomplish those goals, the project needs to take an approach that encourages openness (of process and of end-product), one that reflects diverse and unique users and their context, and one that embraces a realization of a platform that meets the end-user's needs.

The Prosperity4All community understands that they are creating the infrastructure to enable an ecosystem to grow. But they will not control that ecosystem, nor can they predict how it will grow, adapt, evolve, and change.

Understanding the complex
The world has emergent, unpredictable systems

Our world is made up of emergent systems that grow, evolve, change, and adapt over time (Czap et al., 2005). Often these systems are made up of many smaller systems interconnected in a complex that has distributed management, oversight, policies, etc. Prosperity4All is building an inclusive ecosystem, but the builders of that ecosystem cannot produce a precise road map of how these pieces are going to be utilized by different users, it cannot account for all the different contexts those users will be acting within, and it cannot know how the intended ecosystem will emerge and grow over time. This is the complex context within which designers on P4A are creating interfaces.

This predicament is not unique to Prosperity4All. There is a growing articulation of emergent systems across a number of domains. In fact, it is difficult to find a domain that doesn't deal with some level of unpredictability, adaptability, interconnectedness, evolution, and more. Complex systems abound: Igor Nikolic, a chemical and bioprocess engineer at the Delft University of Technology, explains: "complexity is understood by some as the inability of single language or single perspective to describe all the properties of a system we observe" (TEDx Talks, 2010). And when we examine our deeply technological world, most of what we encounter fits this description.

The challenge from the perspective of building these complex systems is to find a way that allows the ultimate system to grow and change and evolve—to emerge. We *want* the system to be unpredictable, extensible, mutable, and adaptive. Another challenge is to find a way to understand how people and systems interact in an emergent, evolving way, and to find a way to create the conditions that enable the emergence to happen inclusively (without creating barriers, creating opportunities, meeting unmet needs, etc.).

People are unpredictable too

There is significant incentive (and effort) to predict human behavior. If we can predict what you need or want, we can tailor ads or products or interfaces to you. If we can predict what you'll like or dislike, we can direct your digital behavior and more. Many online businesses are trying to predict human behavior through increasingly sophisticated algorithms. And those groups are finding the limits of computer algorithms in predicting human activities. "The human mind's algorithms are far more sophisticated than anything Silicon Valley has yet devised, but they're also heavily reliant on heuristics and notoriously prone to folly" (Oremus, 2016).

This human unpredictability has a significant impact on design. If we are only consistent within a given context, then we cannot predict how an individual will interact with a system since their context is ever-changing. None of us is predictable, but we are so eager to understand human behavior as something orderly and stable. Unfortunately we have, uncritically in many ways, come to accept an averaging of behavior in order to understand each other.

Todd Rose's recent book *The End of Average* deals with the history and context in which we culturally came to accept and rely on "averagarianism," "the supposition that individuals can be evaluated, sorted, and managed by comparing them to the average" (Rose and Ogas, 2016). Rose states, "individuals behave, learn, and develop in distinctive ways, showing patterns of variability that are not captured by models based on statistical averages" (Rose et al., 2013). And yet we rely on the average to make decisions that have deep and lasting effects on all of us. We do this "because averagarianism worked better than anything else that was available" (Rose and Ogas, 2016). Rose, among others, is working within the interdisciplinary field "science of the individual," a science that contends, "the way someone behaved always depended on both the individual and the situation" (ibid, 105). The impact of this work is dramatic and far-reaching. Not only does it suggest we need new models and frameworks for understanding human behavior, but along with that the systems and structures that they exist within.

FSG, a social change research firm, notes that "over the last several years, there has been an increasing realization in the social sector that systemic change is not linear, predictable, or controllable. We are learning that social problems are more resilient than previously thought and that traditional means of tackling them often fall short. This is not due to bad intentions, but almost always due to faulty assumptions" (Preskill and Gopal, 2014). We are fundamentally confronted by the following truism: systems and people both evade predictability. In both we can look for the emergence of patterns of behavior or trends or opportunities, but it is fundamentally only in their dynamic interconnection (people with systems) that we can understand the growth and emergence of complex systems.

We know that both people and their contexts are mutable and changeable and unpredictable in many ways. We find that "the external environment is not some kind of pear to be plucked from the tree of external appraisal, but a major and sometimes unpredictable force to be reckoned with" (Mintzberg and Waters, 1985). Pairing an inclusive design perspective with an understanding of the complexity and unpredictability of our world and the systems in it leaves us with real tools to understand how to design for inclusion, openness, and flexibility. Designing for adaptability, for personalization, and for the individual means designing in a way which acknowledges that we can't anticipate and predict—that we must empower users to decide and we must build systems that are flexible and able to adapt.

HEALTH AND HEALTHCARE ARE COMPLEX SYSTEMS
The multisided market possibilities in healthcare

Multisided markets or platforms have been shown to diversify both demand and supply. Previously fragmented demand and supply are connected and aggregated to produce new economies of scale and alternatives to marketing. As demands are connected to supply through the services of the platform, resources previously devoted to marketing can be reallocated to greater customization. Inclusively designed multisided markets have the potential to address needs that are unmet by the current healthcare "marketplace." Often innovation is born from unique needs and desires. As has been shown in technology (Jacobs, n.d.), "the accessibility problems of today are the mainstream breakthroughs of tomorrow. Autocomplete and voice control are two technologies we take for granted now that started as features aimed at helping disabled users use computers, for example" (Brownlee, 2016). Some recognize that with this "reverse inclusion," "the world regularly becomes a better place for everyone when we concentrate on taking the needs of disabled and older people more seriously. And, yes, there is money to be made" (Beyond Inclusion, 2011).

Some users need, some prefer, some want—technology should just enable this range. Markets that enable producers and consumers to customize, personalize, and create one-size-fits-one solutions will lead to greater innovation, will meet currently unmet needs, and will tap unreached markets. The WHO outlines a number of unmet needs in its fact file on health inequalities (WHO 2016).

Current gaps in healthcare occur because of:
- social status, income, ethnicity, gender, disability, or sexual orientation
- limited availability of services (geographic)
- those in vulnerable employment situations
  Some barriers to healthcare are
- cost
- physical barriers
- inadequate skills and knowledge of health workers

An inclusively designed multisided market built as a health ecosystem can begin to address many of these barriers. The combination of an inclusive design process and a multisided market presents the potential to disrupt existing, traditional markets, and address the inherent asymmetries they promote. Multisided markets are opportunities for having a fundamental impact on exclusion by addressing unmet needs and disrupting various industries that exclude.

For continuous improvement, a successful multisided market must encompass a dynamic and well-designed feedback loop. This provides opportunities for all stakeholders or participants to evaluate transactions. This evaluation can be instrumented with metrics or algorithmically gathered performance measures. In this dynamic, continuous evaluation, there is an opportunity to address the needs of individuals excluded by traditional health research, which tends to privilege the hypothetical average. An inclusively designed multisided market would gather and analyze bottom-up or individual data. This can be used to enable self-knowledge on the part of the participant by displaying results in an inclusively designed fashion; and it can be used to adjust the experience of the platform for each individual user, thereby supporting one-size-fits-one. Lastly the results can be aggregated and used to identify current gaps in service.

These disruptive trends are in some cases creating opportunities for addressing needs beyond just the majority and providing conditions where individual needs can be met. These disruptions provide the conditions for producers and consumers to both think of novel, innovative solutions. There are a number of anecdotes that convey this, such as the story of a carpenter in South Africa and a special effects artist and puppeteer from Washington who make and share plans for 3D printed prosthetic limbs, changing the lives of children around the world offering a way to make low-cost prosthetics (Henn and Carpien, 2013). Another example is the story of the

12 year old who built a braille printer out of Legos and dramatically reduced the price of the device (Kooser, 2014). Other examples include collaborations such as the Inclusive Design Research Center and Uber codesign activity, creating an accessible Uber solution piloted in Toronto (https://newsroom.uber.com/canada/ocad/), and the collaboration between Inclusive Design students and Bridgepoint Hospital to create healthcare resources that work for individuals experiencing cognitive decline (Black, 2016).

Where traditional markets are falling short technology is acting as a catalyst empowering entrepreneurs and savvy consumers to disrupt, reconfigure, and reimagine their options in the marketplace. The examples above are small, but represent a shift in healthcare and self-management. In a sector so personal and controlled by complex systems like medical personnel and insurance companies, there are many opportunities for a change in dynamic.

An inclusively designed ecosystem, that connects producers and consumers and lowers the barrier of entry to not only those who are digitally savvy, can have a profound impact on the current gaps in healthcare. The technology exists, we know how to make inclusive websites and applications, we know how to build platforms that encourage collaboration and sharing, we know how to build platforms that adapt and grow and are self-governed. We are at an exciting time in technology and an opportune time in healthcare.

CONCLUSION

Inclusive design is about addressing the unmet needs of individuals in our complex world. And it is about empowering individuals in a complex world made up of complex systems, many of which are exclusionary. Inclusive design makes it possible to disrupt those systems and to break down barriers that currently exist. Inclusive design has the potential to end cycles of exclusion. It requires a perspective shift and an unwavering willingness to take on the hard problems. It requires that we solve for mismatch and provide opportunities for personalization, flexibility, and openness in whatever we design and build. From this approach comes greater inclusion, greater social prosperity, and innovation. One area where this disruption and innovation is primed to have an impact is in the health sector. Healthcare broadly and e-health specifically are complex systems that would benefit from an inclusive design reframing.

ENDNOTES

1. For a thorough discussion of the differences between universal and inclusive design, see "What is Inclusive Design" (Treviranus n.d.).
2. The Inclusive Design Research Center has created Design Guide Cards that help others understand the various principles, practices, tools, and activities that make up inclusive processes. They can be found at the following link: <https://guide.inclusivedesign.ca>.
3. To see strategies for making inclusively designed personas and scenarios refer to the Inclusive Design Research Center's Design Guide Cards <https://guide.inclusivedesign.ca>.

REFERENCES

Ajami, S., Bagheri-Tadi, T., 2013. Barriers for adopting electronic health records (EHRs) by physicians. Acta Inform. Med 21 (2), 129–134. http://doi.org/10.5455/aim.2013.21.129-134.

Ayotte, D., Vass, J., Mitchell, J., Treviranus, J., 2014. Personalizing interfaces using an inclusive design approach. In: Stephanidis, C., Antona, M. (Eds.), Universal Access in Human–Computer Interaction. Design and Development Methods for Universal Access: 8th International Conference, UAHCI 2014, Held as Part of HCI International 2014, Heraklion, Crete, Greece, June 22–27, 2014, Proceedings, Part I. Springer International Publishing, Cham, pp. 191–202.

Beyond Inclusion and Reverse Inclusion: how fully engaging with the needs of disabled and elderly people can turbo-charge innovation and profitability - Hassell Inclusion [WWW Document], 2011. <http://www.hassellinclusion.com/2011/10/beyond-inclusion-and-reverse-inclusion/#reverse-inclusion> (accessed 28.6.16).

Black, I., 2016. Uber and the OCAD Inclusive Design Institute team up to Improve Accessible Transportation in Toronto.

Brownlee, J., 2016. How Designing For Disabled People Is Giving Google An Edge [WWW Document]. Co.Design. <http://www.fastcodesign.com/3060090/how-designing-for-the-disabled-is-giving-google-an-edge> (accessed 28.6.16).

Burns, C.T., 2015. Christy Turlington Burns: Make childbirth safe in U.S. [WWW Document]. CNN. <http://www.cnn.com/2015/12/02/opinions/turlington-maternal-mortality/index.html> (accessed 28.6.16).

Charlton, J.I., 2000. Nothing About Us Without Us: Disability Oppression and Empowerment. University of California Press, Berkeley.

Crossing the Quality Chasm: A New Health System for the 21st Century, 2001. Shaping the Future for Health. Institute of Medicine.

Czap, H., Unland, R., Branki, C., 2005. Self-Organization and Autonomic Informatics (I). IOS Press, Amsterdam.

Epstein, R.M., Street, R.L., 2011. The values and value of patient-centered care. Ann Fam Med. 9, 100–103. http://dx.doi.org/10.1370/afm.1239.

European Disability Strategy 2010–2020: A Renewed Commitment to a Barrier-Free Europe, 2010. Communication from the Commission to the European Parliament, the Council, the European Economic and Social Committee and the Committee of the Regions.

Fluid Project Wiki, 2016a. Fluid Project Wiki. <https://wiki.fluidproject.org/display/fluid/%28Floe%29+User+states+and+contexts> (accessed 8.7.16).

Fluid Project Wiki, 2016b. Fluid Project Wiki. <https://wiki.fluidproject.org/pages/viewpage.action?pageId=80674818> (accessed 8.7.16).

Global Public Inclusive Infrastructure (GPII) Website, 2011. gpii.net. <http://gpii.net/> (accessed 28.6.16).

Global Public Inclusive Infrastructure (GPII) Wiki, 2016. GPII Wiki. <https://wiki.gpii.net/index.php?title=Inclusive_Design_Guidelines&redirect=no> (accessed 8.7.16).

Henn, S., Carpien, C., 2013. 3-D Printer Brings Dexterity to Children with No Fingers [WWW Document]. NPR.org. <http://www.npr.org/sections/health-shots/2013/06/18/191279201/3-d-printer-brings-dexterity-to-children-with-no-fingers> (accessed 28.6.16).

IPFCC About Us [WWW Document], n.d. Institute for Patient- And Family-Centered Care. <http://www.ipfcc.org/about/index.html> (accessed 28.6.16).

Jacobs, S., n.d. The Electronic Curb Cut [WWW Document]. <http://www.accessiblesociety.org/topics/technology/eleccurbcut.htm> (accessed 5.7.16).

Kooser, A., 2014. 12-year-old builds low-cost Lego braille printer [WWW Document]. CNET. <http://www.cnet.com/news/12-year-old-builds-low-cost-lego-braille-printer/> (accessed 28.6.16).

Mintzberg, H., Waters, J.A., 1985. Of strategies, deliberate and emergent. Strateg. Manage. J. 6, 257–272.

Oremus, W., 2016. Who Controls Your Facebook Feed. Slate.

Preskill, H., Gopal, S., 2014. Evaluating Complexity: Propositions for Improving Practice. FSG.

Prosperity4All, n.d. Prosperity4All. <http://www.prosperity4all.eu/> (accessed 07.07.16).

Rigby, M., Georgiou, A., Hyppönen, H., Ammenwerth, E., de Keizer, N., Magrabi, F., Scott, P., 2015. Patient portals as a means of information and communication technology support to patient-centric care coordination – the missing evidence and the challenges of evaluation. Yearb. Med. Inform. 10, 148–159. http://dx.doi.org/10.15265/IY-2015-007.

Rose, L.T., Ogas, O., 2016. In: Sameness, First (Ed.), The End of Average: How We Succeed in a World That Values. HarperCollins Publishers, New York.

Rose, L.T., Rouhani, P., Fischer, K.W., 2013. The science of the individual. Mind Brain Educ. 7, 152–158. http://dx.doi.org/10.1111/mbe.12021.

Shallcross, L., 2015. Online Health Tools Might Not Help The People Who Need It Most [WWW Document]. NPR.org. <http://www.npr.org/sections/health-shots/2015/10/23/450936409/online-health-tools-might-not-help-the-people-who-need-it-most> (accessed 28.6.16).

Showell, C., Turner, P., 2013. The PLU problem: are we designing personal ehealth for people like us? Stud. Health Technol. Inform. 183, 276–280.

TEDMED, 2011. Michael Graves at TEDMED 2011.

TEDx Talks, 2010. TEDxRotterdam - Igor Nikolic - Complex adaptive systems.

Thomas, A., 2012. Growing problem of "digital exclusion" | Poverty and Social Exclusion. Low Incomes Tax Reform Group.

Treviranus, J., n.d. Inclusive Design Research Centre: What is Inclusive Design? [WWW Document]. <http://idrc.ocadu.ca/about-the-idrc/49-resources/online-resources/articles-and-papers/443-wha-tisinclusivedesign> (accessed 28.6.16).

Treviranus, J., 2014. Leveraging the web as a platform for economic inclusion. Behav. Sci. Law 32, 94–103. http://dx.doi.org/10.1002/bsl.2105.

Treviranus, J., 2016. Life-long Learning on the Inclusive Web. W4A'16, April 11–13, 2016, Montreal, Canada ACM 978-1-4503-4138-7/16/04. <http://dx.doi.org/10.1145/2899475.2899476>

Vaughn, J.R., 2009. The Current State of Healthcare for People with Disabilities | NCD.gov [WWW Document]. <https://www.ncd.gov/publications/2009/Sept302009#Health%20and%20Health%20Disparities%20Research> (accessed 28.6.16).

WHO | Disability and health [WWW Document], 2015a. WHO. <http://www.who.int/mediacentre/factsheets/fs352/en/> (accessed 28.6.16).

WHO Statement on Caesarean Section Rates, 2015b. World Health Organization.

WHO | Fact file on health inequities [WWW Document], 2016. WHO. <http://www.who.int/sdhcon-ference/background/news/facts/en/> (accessed 28.6.16).

PART III

Safety and Privacy

CHAPTER 5

Privacy, Trust and Security in Two-Sided Markets

P.S. Ruotsalainen
University of Tampere, Tampere, Finland

INTRODUCTION

The way healthcare services are provided and used is currently undergoing a simultaneous and meaningful paradigmatic and technological change. The new holistic biopsychosocial health model highlights the understanding that diseases are not only caused by some identifiable physical or chemical event. Instead they have biological, psychological, social, and environmental components (Ruotsalainen et al., 2015; Seppälä et al., 2012). This means that healthcare services, such as prevention, early detection of diseases, and disease management, require the availability of personal health information (PHI) that exceeds the content of current electronic health records (EHR). An increasing part of healthcare services, such as in-home care, remote monitoring, and diagnosis as the collection PHI, takes place outside regulated healthcare organizations (hospitals, doctor offices, and healthcare stations). All those services require the availability of modern communication networks.

Parallel to the ongoing changes in healthcare a new concept of ubiquitous health has developed. It uses ubiquitous technology to offer individuals a health service anytime and anywhere. In ubiquitous health the service user can be a person (a consumer) without having the status of the patient (Ruotsalainen et al., 2012). In ubiquitous health users can manage own health without healthcare professionals, and service providers can be nonhealth organizations (Ruotsalainen et al., 2015). Typical services of ubiquitous health are the management of the personal health record (PHR) and the use the information attained from wearable sensors and motes for personal health management and digital health applications.

Technology, especially ICT technology, has been for years a strong driver that changes not only the way healthcare and health services are provided but also health information systems. Computers and networks enable mobility, online information sharing, digital imaging, and the possibility to store all EHRs and PHRs forever. ICT technology also digitalizes both the content of the PHR and, increasingly, health services (e.g., applications for early detection of diseases, digital doctors, and remote

E-Health Two-Sided Markets.
65

monitoring). The possibility of processing digitalized health information and the availability of global online networks have together enabled new healthcare service models such as e-health, digital health, and the existence of ubiquitous health.

E-health and digital health require a service architecture that is customer-centric, dynamic, and multidisciplinary. The current organization of hierarchical and static service architecture used in healthcare cannot be moved without meaningful changes to the distributed and dynamic environment of e-health and ubiquitous health. Approaches such as platforms, multisided markets, and ecosystems have been successfully used for years in e-commerce (e.g., Google, Amazon, eBay, PayPal). Economists and researchers have also proposed the model of multisided markets for e-health and ubiquitous health. The acceptance of the multisided market model in healthcare changes the way health services are selected and used. It also changes the service provider–customer relationship. To be desirable for customers, the multisided e-health market as whole should be ethically acceptable and trusted, with conformance with regulations and laws, it must be secure, and it must offer the customer protection of his or her information privacy.

E-health, digital health, and ubiquitous health increasingly take place in the distributed, cross-jurisdictional and insecure information space (Ruotsalainen et al., 2013). This environment generates many privacy threats which do not exist in current healthcare information systems, such as: there is no predefined trust between stakeholders, ethical codes, privacy features, and business models of service providers are often unknown, and it is difficult for the service user to know which laws the service provider applies (Ruotsalainen et al., 2012). In health ecosystems customers and patients are not anymore only passive objects. Instead they are active; they make personal choices based on rich information about service quality and trustworthiness of service providers. They also want to control how their PHI is collected, shared, disclosed, and postreleased. All those changes have a meaningful impact on how security, privacy, and trust are understood, created, and managed in e-health ecosystems.

Privacy and trust are the social code glue between persons, and inside a society. Privacy is also a human right and without trust the health society will hardly work. Trustworthiness is a code between a person and a business (e.g., between a patient and a health service provider). Ethics, human rights, ethical codes, rules for fair information processing and regulations together determine how health service professionals, ICT developers, and health information professionals perform their work. By defining concepts of right and wrong, ethical codes give us answers to which actions are right or wrong.

Information ethics tries to give answers how information can be ethically collected, used, and disseminated in society. Computer ethics addresses the ethical issues and constraints that arise from the use of computers. In the modern information society the role of information and computer ethics is rapidly increasing in prominence.

In healthcare, medical privacy has been a cornerstone of the patient–doctor relationship since Hippocratic times. It is an implied agreement of keeping information about a patient confidential.

In single markets ethical issues and privacy requirements are well defined (e.g., in healthcare there exist widely accepted domain-specific ethical codes, privacy laws, norms, and international standards). In spite of this, the e-health ecosystem and two-sided markets take place in an unsecure information space (i.e., fully or partially outside the traditional healthcare domain) where many services are offered by unregulated health providers, and there has been until now little or no discussion about ethical challenges, privacy, and trust issues associated with this new service environment. To be successful, it is evident that e-health ecosystems must be trustworthy and must protect service user's information privacy. Because current static and predefined trust and privacy models used in healthcare will not work in the environment of the e-health ecosystem, a new holistic framework is needed (Ruotsalainen et al., 2012; Ruotsalainen et al., 2013).

In this chapter, security privacy and trust challenges of multisided e-health markets are analyzed, and a conceptual framework model is developed. With the help of the developed framework model and security privacy and trust analysis, ethical, trust, privacy and security principles, and requirements for tamper proof and trustworthy two-sided e-health markets are proposed.

CONCEPTS AND DEFINITIONS

The concept of e-health shows a significant variability in its scope and definitions. Differences in the e-health concept rise from how health and healthcare are understood. The term healthcare means the medical, nursing, and health professions services such as prevention, treatment, care, and management of illness offered by medical, nursing, and allied health professionals (Digital healthcare, 2006). By the WHO's definition health is a wider concept than healthcare. It means the ability to cope with everyday activities, physical fitness, and high quality of life.

Most definitions conceptualize e-health as "a broad range of medical informatics applications for facilitating the management and delivery of healthcare" (Margulis, 2003). The WHO has defined that "e-health is the transfer and exchange of health information between health consumer, subject of care, health professional, researcher, and stakeholders" having right to use information using communication networks (e.g., the Internet, mobile networks, social networks), and the delivery of digital health services using networks both at a distance and locally (World Health Organization). This definition covers not only regulated healthcare but also ubiquitous health and nonregulated health services. In this book chapter the WHO definition is used.

The following list illustrates typical e-health applications (Telecommunication standardization sector of ITU, 2014):

- The storage and exchange of PHI
- Dissemination of health-related information
- Consumer health informatics
- Sharing and distribution of the content of EHRs and personal health data between professionals, patients, and researchers
- Monitoring of patients' vital signs
- Personal health management using sensors
- Knowledge distribution
- Supporting independent living
- Virtual doctor and hospital services
- Use of mobile devised in collection of PHI
- PHR management
- In-home care services
- E-learning
- Assisted living services
- Health education

The core idea behind the concept of digital health is reshaping health through data and with the help ICT technology. Digital health has been understood as the convergence of the digital technology and genomic revolutions that combines molecular and genomic data, anatomical, physiological, environmental, and behavioral data together. It uses ICT technologies to improve health, healthcare services, and wellness for individuals and populations (Kostkova, 2015). Digital health tries to connect all available technologies such as m-health, big data, EHRs, cloud computing, wearable computing, gamification, e-patients, self-monitoring and personalized health, digital patient applications, and avatar doctors. Digital health is a multidisciplinary domain and its stakeholders include medical professionals, researchers and scientists, engineers, social sciences, public health, and health economics.

Similarly to e-health, privacy and trust are dynamic and context-dependent concepts. A big challenge is that in different societies and countries privacy and trust are understood differently. In computer networks (e.g., the Internet and cloud computing) privacy and trust are increasingly one of the most concerning obstacles (Noor and Sheng, 2011). Without understanding the privacy and trust challenges that exist in two-sided e-health markets, it is impossible create a trustworthy e-health ecosystem.

Information privacy is a personal, elastic, and context-dependent concept. It was originally defined to cover interpersonal communication. In the information society (and in e-health), privacy should cover person-to-computer, computer-to-computer, and organization-to-organization communication (Margulis, 2003). There is no globally accepted privacy model used in information space. Most common privacy models are Westin's information privacy theory, Altman's privacy model, and Pertino's communication privacy management theory. Westin's model addresses how people protect themselves by limiting access of others to themselves (when, how, and to what extent information about them is communicated to others) (Social and Political Dimensions of Privacy, 2003). Altman's model defines that privacy is the selective control of access to the self. In this model, privacy is a dynamic process of interpersonal boundary control (Margulis, 2003). In Pertino's communication privacy management theory (CPM) private information is defined in terms of ownership and therefore individuals have the right to control the distribution process (Petronio and Jennifer Reierson, 2009). In CPM, privacy boundaries can range from complete openness to being completely closed, and rule-based privacy management system is used to regulate the degree of boundary permeability (Margulis, 2003). The newest privacy model is contextual privacy presented by Nissenbaum. It is focused to the question of how to protect privacy online in context. In it, norms determine what types of information are/are not appropriate for a given context, and norms of distribution determine the principles governing distribution (Nissenbaum, 2009).

Privacy level of an information system (e.g., the e-health ICT system) in a context depends on legislation and norms, market features, the nature of information and its sensitivity, characteristics of information user, activities expected, the level of trust of the service provider, technical architecture of service provider's ITC system, and expected benefits for the user (Ruotsalainen et al., 2013; Lederer et al., 2002). All of them should be taken into account in the development of trusted and tamper proof two-sided e-health Market.

Regulatory and self-regulation-based privacy models are most common in healthcare information systems. In regulatory privacy models rules are based on nationally/internationally accepted laws. This model is widely used in healthcare. In healthcare, the dominant approach to address privacy is a combination of transparency and choice i.e., notice-and-consent, or informed consent (Margulis, 2003). It is widely used in healthcare to enable the patient to control the disclosure of the content of his or her EHR. Unfortunately, in e-commerce a common practice is to enable customer choice to mean a "take-it-or-leave-it" decision. Self-regulation is based on service provider's ethics and morals, on business needs, rules, and norms. This model is widely used by the industry.

The contextual privacy approach assumes protection of privacy by norms of a specific context, and rules control the appropriate information flow. Contexts are social settings characterized by roles, relationships, norms, enforcement mechanisms, and internal values (examples of context: healthcare, education, family). Based on Nissenbaum, in networked and distributed information systems informed consent is failed, opt-out is unfair, and privacy based only on service providers' ethical norms is insufficient (Nissenbaum, 2009).

Information security means protecting both information and information systems from unauthorized access, use, disclosure, disruption, modification, or destruction. Its elements are accountability, availability, confidentiality, and integrity (often nonrepudiation is also added) (International Organization for Standardization (ISO), 2008). Confidentiality is about identifiable personal information (PII). It is an agreement about how the data will be managed and how it is access controlled by the data controller or processor. Confidentiality means that the entity processing PII has the responsibility to protect data against misuse and unauthorized use.

Trust is a complex, context-dependent, and dynamic concept. It can be understood as the subjective probability that a system will perform an action before a person can monitor it (Ruotsalainen et al., 2013; Gambetta, 1988). Most common trust models are belief, organizational, dispositional trust, direct trust, recommended trust, and computational trust. Ruotsalainen et al. have noted that in ubiquitous environments belief and organizational trust is ineffective and recommended trust is in most cases only a summary of opinions of unknown entities. Therefore, computational trust should be used to create trust in information spaces. Trust can be calculated using trust attributes, such as ability, willingness, predictability, privacy policy, and system properties, and information got from external monitoring and previous privacy history.

Privacy and trust are interrelated in such a way that increased levels of trustworthiness reduce the need for privacy (Ruotsalainen et al., 2012). For example, if the e-health customer trusts in the service provider and its information system, the customer can accept to use services without specific privacy protection needs.

REGULATED HEALTHCARE AND TWO-SIDED E-HEALTH MARKETS

Traditionally, healthcare services have been organized using the one-side market approach. An increasing number of hospitals, primary care centers and healthcare stations have started to offer e-health service for patients (e.g., monitoring of patients' vital signs in the home, patient–doctor communication, and health education services). The focus of those solutions is typically professionals' needs and service organizations' efficiency, and they are typically part of a service organization's legacy system (e.g., part of Hospital EHR-system or the Regional Information System). A paternalistic patient role is also widely assumed (i.e., the user is the object of care, information source, and passive object for clinical decisions). E-health services offered by healthcare seldom or never enable the patient to select services based on personal needs and quality. Furthermore, in those applications the patient cannot define their own privacy rules concerning the use and disclosure of his or her PHI. This kind of implementation of e-health implicates that there is no real service paradigm change.

Outside the regulated healthcare domain, many nonregulated health service providers associations and commercial application providers offer e-health services in the

Internet and on mobile phones (those services are sometimes called "wellness management services"). Nonregulated e-health solutions seem increasingly adapted the idea of software as service (i.e., a model a third party hosts an application and makes them available over the Internet or mobile network).

In their original form, two-sided markets, also called two-sided networks, are economic platforms having two distinct user groups that provide each other with network benefits. The intermediate that creates value primarily by enabling interactions between two (or more) customers is called a platform. In two-sided e-health market, service users (customers or patients) can select services from competing providers using some criteria such as trustworthiness or quality of services.

There are many business models for two-sided markets such has freemium (basic services are offered for free while others are charged) and "As a Service" (only the usage of service is charged) (Mettler and Eurich, 2012). In e-health, it is typical that one party (e.g., Insurance Institute) has high interest in subsidizing customer's access to services (Mettler and Eurich, 2012). The original economical idea of two-sided markets has been also extended. For example, two-sided markets have been understood as "a meeting place for two sets of agents who interact through an intermediary or platform," i.e., the platform is an audience for users.

Two-sided market theory is a nice theory that tells us how the platform works economically. Most research of multisided markets is focused to competition, pricing, and network externalities. In this chapter, the focus is on how customers using e-health services through two-sided markets can trust in it, and how acceptable levels of security and privacy can be reached. It is also assumed that in e-health two-sided markets consumers and providers act directly via the platform, and the platform offers services and tools for interactions (Heitkoetter et al., 2012). If needed the platform also orchestrates customer's services.

SECURITY, PRIVACY, AND TRUST IN HEALTHCARE AND TWO-SIDED E-HEALTH MARKETS

It is widely understood that security, privacy, and safety requirements of healthcare products significantly differ from requirements set for traditional consumer products. Similarly, patient–doctor relationships differ meaningfully from user-producer relationships. The patient–doctor relationship is a fiduciary relationship, and it is based on mutual understanding. Patient rights are legally expressed (e.g., in the Act of Patient Rights). From a security and privacy point of view the patient has the right to expect to receive correct information, and that his or her information privacy is guaranteed.

In healthcare, codes for medical ethics and rules for fair information processing form a global framework for the collection and use of patient information. The healthcare service provider has a responsibility to maintain the confidentiality of patient information (Ruotsalainen et al., 2015). Depending on regulations in force, the patient can use personal information autonomy (i.e., consent/opt-out) and control the disclosure of the content of his or her health record.

Healthcare has long been an institutionalized and regulated organization in a controlled and secure environment (Ruotsalainen et al., 2015). In healthcare, it has been traditionally assumed that patients blindly trust in service providers' ethics, knowledge, and decisions, and that the caring persons are qualified. Furthermore, it is widely expected that patients have predefined dispositional or organizational trust, i.e., patients are confident that information systems (e.g., the EHR-system) service providers use are secure, fulfill regulatory and legal requirements, and that patient's PHI is protected against unauthorized access and misuse. Healthcare information systems (e.g., hospital information system) have been until now security oriented and privacy has not been a concern. Most of today's healthcare information systems are focused on security management and access control. These assumptions are not valid in e-health digital health and ubiquitous health because the collection and processing of PHI take place over organizational geographical and regulatory borders. Similarly, security-based authentication and role-based access control approaches cannot guarantee security in cross-domain PHI sharing in information space (Ruotsalainen et al., 2013).

In the case of two-sided markets, in services used and produced inside the same jurisdiction, the same domain-specific security and privacy protection laws and norms apply. In the case of cross-jurisdictional services used via the two-sided markets, the situation is more complicated because healthcare specific security and privacy laws and norms differ. As a minimum, the user should be informed which regulations the service provider is using. Based on this information, the user can define own security and privacy protection requirements before he or she starts to use services, e.g., define a requirement that the service provider processing the customer's PHI must follow the regulations of the customer's own home domain.

Table 5.1 shows that there are differences both in security and privacy principles and in trust models used in healthcare and what is expected for e-health two-sided markets. For example, the principles of medical ethics are not automatically used by nonregulated health service providers (International Medical Informatics Association (IMIA)). In healthcare, personal privacy policies are seldom or never supported, and in e-health two-sided markets predefined organizational trust does not exist. Concerning security, in e-health two-sided markets no-repudiation of customers and service providers cannot automatically be expected. This all indicates that privacy and trust models and rules used widely in current healthcare cannot be moved in the form that they stand in to environment of two-sided e-health markets.

It is widely accepted that PHI information must be collected and processed ethically and fairly. These requirements are global and technology independent, i.e., also the two-sided e-health markets must apply them. Up to now, there is a lack of

Table 5.1 Ethical principles, privacy and trust in health care and two-sided markets

Principle	Healthcare services used inside a closed domain	Health care services used through two-sided markets	Health services used through two-sided markets
Ethics principles and values	– Principles of Medical ethics – Fair Information processing principles – Ethical codes for health information professionals	– Principles of Medical ethics – Fair Information processing principles – Ethical codes for health information professionals	– General ethical principles – Fair Information Processing Principles
Trust	– Predefined and static organization trust at jurisdictional level	– Agreements based on trust – Trust and privacy certificates	– Trust is based on measured features of service providers – Every service provider has own trust level
Security	– Confidentiality availability, accountability and integrity – Standardised security services	– Confidentiality availability, accountability and integrity – Use of standardised security services	– Confidentiality availability, accountability and integrity – Measured security services – Non-repudiation
Information Privacy	– Regulatory privacy model – Privacy by security services – Informed consent	– Contextual privacy – Privacy contract – Gross-organisational privacy agreement – Informed consent	– Contextual privacy – Regulations and contextual norms – Personal privacy policies

common security, privacy, and trustworthiness principles and rules which all stakeholders of the two-sided e-health markets accept, and there is an urgent need for a harmonized ethical framework and common privacy principles for all actors in the two-sided e-health market to apply.

CONCEPTUAL MODEL FOR TWO-SIDED E-HEALTH PLATFORM

The term platform has many meanings. It has been used both in an abstract and a technological sense. In computer science it means a combination of hardware and software modules and operating systems (e.g., the Internet or a website). According to Rochet and Tirole the term platform is an abstract idea that characterizes services,

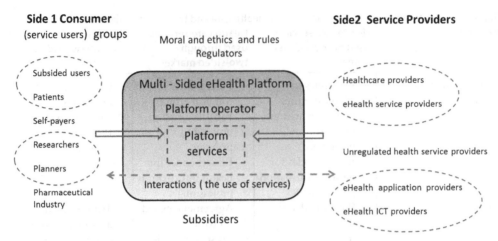

Figure 5.1 Conceptual model for a two-sided e-health platform.

products, and firms that mediate transactions between two or more groups of agents (Gawer). Beyeler et al. have presented a more practical definition, i.e., "A platform is a collection of equipment, facilities, and standards that facilitates a particular kind of interaction" (Beyeler et al.).

In the two-sided e-health market the platform is a computer platform that manages relations and information flow in the surrounding e-health ecosystem. It acts as a mediator or broker between the market's sides. A conceptual model for a basic e-health two-sided platform is shown in Fig. 5.1. In it, the platform is a real-life platform where customers use e-health services, but billing, payment, and membership mechanisms of the two-sided markets are not included. The conceptual model in Fig. 5.1 follows the canonical model of two-sided markets presented by Roche and Tirole (Rochet and Tirole, 2004). This basic model can be called a direct-to-consumer model (de Reuver and Keijzer-Broers).

The platform is surrounded by service provides and consumers. On the consumer side are subsided users (e.g., patients and employees), individuals paying for services directly by themselves, and secondary users of PHI (individuals, groups, and industrial partners). On the service provider side of the platform are three categories of service provider:

- Regulated healthcare service providers that provide healthcare services for selected patients as well as general health information for all customers (e.g., health management and lifestyle education information)
- Providers offering nonregulated e-health services for all customers
- ICT service providers (e.g., cloud services for PHI retention) and e-health application providers

Around the platform there are different distinct groups of customers such as (Fig. 5.1):

- Healthy persons searching preventive e-health services and persons managing their health (e.g., a chronic disease) without the help of healthcare professionals
- Patients receiving subsided healthcare services via the platform
- Secondary users such as researchers and planners
- Third party users such as the pharmaceutical industry

This all means that the platform in Fig. 5.1 can be seen as a multisided e-health platform.

The platform operator (e.g., a Tele-operator) has a strong role because it has the responsibility to implement security and privacy management principles and rules in practice (Beyeler et al.). The platform operator must also guarantee the availability of platform services.

The subsidiser is a third party (e.g., the government, insurance organization, or foundation). Such kinds of third party can be also located between the service used and the platform (de Reuver and Keijzer-Broers). The subsidizing party can control which providers can be members of the market and which customers are allowed to use the services it subsidizes.

Regulators and service subsidizers are not directly using the platform's services. Regulators are either formal, such as the government, or associations, such as WHO, OECD, or a standardization organization. Governmental regulators set laws, degrees, and mandatory norms. Associations publish guidelines such as the Guideline for Fair Information Processing, ethical codes and health informatics standards, which together regulate security and privacy and information processing functionalities of the platform. Some regulations such as international privacy protection directives are mandatorily to be applied by the platform and by all service providers. The platform operator can also define their own rules, such as privacy policies, which service providers and users of the platform must follow. In such a way the platform works as an internal regulator and can effectively control the use of services and interaction through it (Boudreau and Hagiu). Because the e-health multisided market relays both healthcare and health services and the communication between customers and service providers takes place through the platform, the platform must support customer protection, patient safety, information security, and different privacy policies.

The role of the platform is to be a mediator and coordinator between service users and providers, i.e., it balances different interests of consumer and service provider groups (Evans, 2003). Generally, a platform can both relay physical products and provide services (Eisenmann et al.). The platform in Fig. 5.1 relays e-health and digital health services and PHI related to the use of those services. In spite of users and service providers interacting (i.e., customers and patients using services) through the platform, the actual e-health service application is not part of the platform. Platform providers can be service providers (e.g., a hospital) or an independent organization such as a telecommunication organization or an association.

The platform can be also part of a wider e-health ecosystem (Mettler and Eurich, 2012). In an e-health ecosystem there can be many different e-health platforms, e.g., one owned by a healthcare management organization (HMO) offering in-home care service, monitoring of patient's vital signs and patient education, and another owned by a nonprofit association supporting independent living and personal health management services. Services can be either competing or supplementing, and the customer/patient can select a service combination that meets his or her personal needs (i.e., the ecosystem forms a multihoming e-health market).

Basic services of the platform include integration services, interaction and communication services, security services, privacy protection services, mediation/negotiation services, and booking and billing services. Interaction services require that the platform has reasonable communication interfaces which enable customers and providers easy plug-and-play connection. The platform can also offer application interfaces (API) to software development tools used by service providers.

Openness is the basic assumption for multisided markets, i.e., customers can select any service provider and any service available. In healthcare it is typical that the subsidizing organization (insurance organization, employer, or foundation) is the policy holder concerning which service providers and services an insured patient/person can select. Therefore, the platform should also support service distribution policies defined by subsidizers.

Service providers are certified healthcare service providers (e.g., a certified medical doctor or nurse), healthcare institutions or nonregulated persons/organizations. In the case of healthcare services, the service user has a patient–doctor relationship but the use of a nonregulated health service means that the user has the role of customer. A customer or patient can have both roles at the same time. Providers can offer their own e-health services or sell third party services under their control.

The platform owner can be an insurance institute, a health management organization (HMO), an association tele-operator. The platform can be also an application service (SaaS).

Users and service providers use the platform via communication channels such as the Internet and mobile networks. The network around the marketplace is legally independent of the platform, but it should be secure and protect information privacy (de Reuver and Keijzer-Broers).

From regulatory points of view, the e-health multisided market can have different service provider–customer combinations, such as:

- The platform and its users are located inside the same legal and regulatory framework (e.g., inside a country). Services provided and used are healthcare services (e.g., in-home patient monitoring and management of chronic diseases)

- The platform and customers are located inside the same legal and regulatory framework, but services provided and used are a mixture of healthcare and nonregulated health services
- Services the platform offers are regulated healthcare services, but service providers and users can locate anywhere in the information place
- Services the platform offers are a mixture of healthcare and nonregulated health services, and service providers and customers can locate anywhere in the information place

Each combination has its own ethical security privacy and trust challenges are discussed in detail in the following parts of this chapter.

SECURITY PRIVACY REQUIREMENTS FOR THE E-HEALTH MULTISIDED PLATFORM

Technically it is possible to build a closed and fully secure multisided e-health market solution, but is economically and functionally unrealistic because increasingly e-health and digital health do not take place inside a closed, controlled and secure domain where services are only provided by local healthcare organizations (a hospital, primacy care center, or regional/national health management organization). Modern e-health and digital health takes place in a global information space. Even in the case that all e-health services are produced and used locally, the multisided platform is typically located on the Internet and many of its applications use global ICT services such as the Cloud. This means that the communication between the platform, consumers, and providers takes place in an insecure environment such as the Internet and mobile networks. From this point of view, the multisided platforms and the ubiquitous health have many common features and concerns, i.e., many of the security and privacy challenges that exist in ubiquitous health are relevant also for the multisided e-health markets.

The following security and privacy challenges existing in ubiquitous health systems are relevant for multisided e-health markets (Ruotsalainen et al., 2012; Ruotsalainen et al., 2013):

- There is no predefined trust between customers, the platform, and service providers
- The customer cannot know in advance the security and privacy status of the platform and service providers. Therefore he or she cannot make information and trust based privacy decisions
- Primary and secondary use and postrelease of PHI is difficult or even impossible to control by the customer
- Service providers often collect more information than the customer is aware of, and the customer cannot control this collection

- It is difficult or impossible to know in which jurisdiction the platform and service provider are located, and which laws and regulations they apply
- Security and privacy regulations differ per country and the customer cannot define personal context-aware policies (rules)
- Healthcare specific laws and rules regulate only the use of patient's health information inside a specific healthcare domain. Those laws and rules do not regulate e-health service and the processing of PHI that takes place outside the regulated healthcare
- The customer cannot control how and for what purposes the platform and service providers use his or her PHI and to whom they disclose it
- Service providers and the platform lack openness and transparency concerning their trust level, privacy status, and security features

It is inadequate if the multisided e-health platform manager and service providers only publish their own security and privacy polices (i.e., a security manifesto) and expect that the customer blindly trust on them. Instead, it is necessary that the real-life implementation of the platform is certified to be trusted and secure and that it protects consumer's information privacy based on his or her personal policies. Furthermore, it is necessary that the way the platform and service providers process customer's PHI is compliant with laws of consumer's home country. The general requirement is that PHI processed by service providers and communicated through the platform is protected against unauthorized use and misuse, secondary use, and unauthorized postrelease. In the next part of this chapter, more detailed requirements are presented.

THE WAY TO SECURE AND TRUSTWORTHY MULTISIDED E-HEALTH MARKET

Trustworthiness is the enabler for a successful multisided e-health market. Without it the customers do not dare to use the platform and its services. Trustworthiness requires that the whole information processing system, i.e., the platform, service providers, and communication networks used are trusted, and that PHI is collected, processed, and stored ethically, legally, and in line with policies set by the customer.

For a trustworthy multisided e-health market we need:

- A comprehensive ethical framework with ethical principles, codes, and rules that all stakeholders of the e-health multisided market apply
- A privacy model with fair information processing rules that enable the customer to define personal privacy policies
- Rules and policies which take into account security and privacy concerns existing in environment of the e-health ecosystem (i.e., the information space)
- Openness and transparency that enable the customer to make information-based decisions concerning the use of the platform and the selection of services

The e-health market must have an ethical framework that is based on the following ethical principles, values, and codes:

- General ethical principles such as freedom, privacy, and principle of transparency (Regulation (EC) No 1982/2006 of the European Parliament)
- General ethical values such as information autonomy, confidentiality, and nonmaleficence
- General principles of information ethics, i.e., information privacy and disposition, openness, security, legitimate infringement, accountability, and principle of access
- Fair information practice principles (United States Federal Trade Commission)
- Ethical code for health information professionals (IMIA)
- ACM/IEEE-CS Software engineering code of ethics and professional practices

Those principles, values, and codes should form the ethical base for multisided e-health markets.

Privacy Model for Multisided E-health Markets

The regulatory privacy model is widely used in healthcare whereas e-commerce widely uses self-regulation. The regulatory privacy model is static and typically focused toward confidentiality. In real life, this approach is often realized by implementing security services (i.e., privacy by security approaches). This means that the regulation-based privacy model alone is too clumsy for multisided e-health markets.

Self-regulation model gives service providers too much freedom to define and change privacy rules based on their own business needs. In real life, it leads to situations where the service provider publishes only a privacy policy notification, and expects that the customer blindly accepts it as a whole (i.e., the customer has only a "take it as it is or leave it" situation) (Ruotsalainen et al., 2013). In multisided markets, it is common that service providers are located in different contexts that have different privacy models, laws, and rules. Therefore, the platform should support contextual privacy rules.

A good solution for multisided e-health markets is a combination of common ethical principles, global regulations, contextual privacy models, and customers' right to define their own privacy policies for any service provider and any context. As discussed earlier, the type of the service provider and the jurisdiction in which it is located impact on the combination of ethical principles and privacy principles it applies (Table 5.2).

Openness of information, trust features and privacy state of the service provider enable the customer to make information-based decisions concerning the willingness to use services. The availability of rich information concerning service providers' and platform's security, privacy, and trust state enables trust calculations and monitoring of service providers privacy levels. Based on those data the customer or the patient can define contextual and personal privacy policies (if needed). A summary of information needed for the defining the trustworthiness of the platform and the service providers and for defining contextual privacy policies is shown in Table 5.3.

Table 5.2 Impact of the regulatory framework and the type of service to principles and rules

Common regulatory framework and only healthcare services offered	Common regulatory framework both healthcare and non-regulated health services offered	Cross-jurisdictional healthcare services offered	Cross-jurisdictional health care and non-regulated eHealth services offered
– General Ethical Values and principles of information ethics – Ethical Code for Health Information professionals (IMIA) – Fair information processing rules (FIPP)	– For healthcare: same as in column 1 – For health services: General ethical values and principles of information ethics – Fair information processing rules (FIPP)	– General Ethical Values and principles of information ethics: – Ethical Code for Health Information professionals IMIA) – Fair information processing rules (FIPP)	– For healthcare services: rules shown in column 1 – For health services: General Ethical Values and principles of information ethics – Fair information processing rules (FIPP)
– Predefined organizational trust	– For healthcare services: Predefined organizational trust – For health services: measured trust based on reliable information	– Trust is defined in the service agreement or the trust certificate	– For healthcare services: Measured trust or it is defined in the service agreement – For other services: Measured trust based on reliable information
– Standardized services for confidentiality, availability, integrity and accountability	– Standardized services for confidentiality, availability, integrity accountability and non-repudiation	– Standardized services for confidentiality, availability, integrity and accountability	– Standardized services for confidentiality, availability, integrity and accountability and non-repudiation
– Privacy rules are predefined by domain specific health care specific laws	– For healthcare services: Privacy rules are predefined by domain specific health care specific laws – For health services: Contextual privacy and customer's privacy policies	– Harmonized predefined privacy rules or rules are negotiated and agreed in the service agreement	– For healthcare services: harmonized predefined privacy rules or rules are negotiated and agreed in the service agreement – For health services: Contextual privacy, customer's privacy policy

Table 5.3 Information needed for the definition of the level of trust of stakeholders in the multisided e-health markets

The platform	Which ethical values and rules does the platform apply?
	What is the jurisdiction in which the platform is located?
	Which laws, norms, and standards does the platform apply?
	What is the privacy model of the platform?
	Which security and privacy polices does the platform use?
	What is the security architecture of the platform?
	Is the platform certified by security and privacy, and if so, by whom?
	What information does the platform collect (including metadata), and for what purposes are the data used?
	To whom does the platform disclose PHI?
	Which local rules are established by the platform?
	What kind of agreements concerning the use of PHI exists between the platform and the service providers?
	How does the platform balance conflicting privacy and security interests?
	Which are the implemented security and privacy services and safeguards?
	How does the platform monitor user?
	How does the platform prevent unauthorized use of its services?
	How does the platform detect misuse of service and the PHI?
	How does the platform prevent the existence of malicious service providers?
	How does the platform identify customers and providers?
	How are customer security and privacy policies are implemented?
	In what way are the rules defined by the platform negotiated and accepted by others?
	How does the platform identify malicious e-health applications?
	How is stealing of user identity is prohibited?
	Who is the owner of the platform (e.g., public or private healthcare organization, nonprofit association, tele-operator, commercial platform vendor)?
	How does the customer join and leave the ecosystem?
The customer	How can the customer monitor the trustworthiness and privacy level of the platform?
	What kinds of tools exist for trust calculation or for defining the level of trust?
	How the customer is informed of misuse and privacy breaches?
	How the customer is informed of malicious service providers?
	How does the customer define their own privacy policies?
	What kind of decision-making tools does the platform offer?
	How does the service user know when he or she has patient status?

(Continued)

Table 5.3 Information needed for the definition of the level of trust of stakeholders in the multisided e-health markets Continued

The service provider	Which ethical values and rules does the provider apply?
	What is the jurisdiction in which it is located?
	Which laws, norms, and standards does it apply?
	For what purposes is customer PHI used, and by whom?
	Is PHI disclosed or postreleased to third parties?
	How is the customer identified?
	How are internal users of the PHI authorized?
	What kinds of security services are implemented?
	How long is customer PHI stored and where?
	How customer PHI is protected against unauthorized access?
	Are there third parties participating in the providing of services?
Communication and interaction services	How are the availability of the network and its services guaranteed?
	How are the information security and privacy guaranteed during communication?
	Is the nonrepudiation of users verified?
	What are the measures against eavesdropping, interruption, hijacking, and interference on communication and data tampering?

SECURITY AND PRIVACY SERVICES FOR TRUSTWORTHY MULTISIDED E-HEALTH MARKETS

The next step after the selection of trust and privacy principles and models and the definition of trust, privacy, and security requirements for the multisided e-health platform is the implementation. This is typically done by the selecting of security tools and privacy protection and enchanting safeguards. Selection of tools and safeguards is typically done by first defining the security and privacy threats, risks, and vulnerabilities of the information system (International Organization for Standardization (ISO), 2008; International Organization for Standardization ISO, 2016; Blobel, 2002). Risk is typically understood as a probability. For the definition of security risks there are standardized methods but privacy risks of networked and distributed information systems are difficult to estimate. Besides risk analysis there are other methods such as data abuse scenario, user activity methods, and modeling and simulation of attacks. Whichever method or combination of methods is used for selection of security and privacy safeguards, it is necessary to understand that the use of technical safeguards cannot alone guarantee success. This is because the multisided e-health market is a socio-technical system and because security and privacy have meaningful human and organizational dimensions (International Organization for Standardization (ISO), 2008).

The combination of security tools and privacy safeguards that a specific multisided e-health system needs depends on legislation and norms, features of the platform, technology used, nature of personal information disclosed, contextual features, information sensitivity, and characteristics of information users (consumers and service providers) (Ruotsalainen et al., 2013; Lederer et al., 2002). Therefore, the following lines present only a basic list of security, privacy, and trust services needed to make a multi-sided e-health platform and its operating environment trustworthy.

Privacy by design is a widely accepted method used in the designing of information systems which collect and process sensitive personal information (Cavoukian). At a practical level, ISO 27000 series of information security management standards form a good base for selecting security and privacy services. ISO/IEC 27799 Health Informatics-Information security management in health using ISO/IEC 27002 is nicely focused to health information and should be used (International Organization for Standardization (ISO), 2008; International Organization for Standardization (ISO), 2016) by e-health ecosystems.

Basic trust, security and privacy service and function for stakeholders of the multi-sided e-health platform are:

1. The platform
   - Identification of customers
   - Definition of customers' roles (both patient and customer roles are possible at the same time)
   - Service that manages log-in and exit to the platform
   - Profiling application for customers (International Telecommunication Union ITU, 2014)
   - Tools to check the nonrepudiation of service providers and customers
   - Trust level calculation of platform (Ruotsalainen et al., 2013)
   - Authentication, authorization, and access control services
   - Audit-trail management and services for the customer to access audit-logs
   - A service for publishing service provider's ethical codes, security, and privacy policies and platform rules
   - Time stamping service
   - Service for the customer to select a service provider or a service based on trust, quality, and service benefits
   - Service for the creation of privacy policies. It enables the customer to define their own privacy policies to control the purpose of primary use, secondary use, and postrelease of PHI
   - Service to recognize malicious providers and customers
   - Service to measure customer satisfaction
   - Service to reject low-quality providers attempts to join in
   - Services to guarantee continuity of services (e.g., backup or nonstop technology such as replication)

- Service to make available service providers' medical qualifications
- Communication management service (e.g., securing that right customer and service provider are linked together)
- Fault recovery service
- Services and tools against malicious code and viruses

2. **The customer or the patient**

   In the case the service provider or subsidizer gives to the patient or customer necessary ICT resources and tools to join to the platform and use its services, they have the responsibility to implement security and privacy protection tools. Outside this, it is widely expected that the customer is responsible for the security and privacy features of his or her ICT installation. As a minimum, it is necessary to protect customers' ICT equipment against malicious codes and viruses.

3. **E-health service providers**
   - Identification of the platform
   - Identification of customers (if needed)
   - A service for calculating the trust level of the platform
   - Log-in and exit service for customers
   - Authentication and authorization of customers
   - Role-based access control services for internal users
   - Services to make available ethical rules and guidelines for the provider to apply, and the security and privacy policies it applies
   - Service to encrypt customer's PHI
   - Encrypted communication services such as SSL
   - Trusted browser service for customers. It should be isolated from providers of other applications
   - Service for audit-trail management
   - Trusted retention service for PHI

4. **Communication networks**

   The communication network used in the e-health ecosystem should enable strong encryption of the PHI during communication. Availability of networks and nonrepudiation of users should be also guaranteed. The level of availability and speed of communication needed depends on the type of e-health application used (e.g., online monitoring of patients' vital signs requires much higher availability and speed than education).

   It is widely accepted that information systems which collect, process, and communicate PHI should be at minimum assessed by trust, security, and privacy. Certification made by an external third party is the best solution. As a minimum, the following features should be checked during the assessment or certification:
   - How ethical principles, values, and codes, and fair information processing guidelines are implemented in practice?

- To what extent is the ICT system in compliance with laws norms and standards of the customer's home country?
- Are implemented security and privacy safeguards and tools in line with expected risk and threats?
- In the case of healthcare services, it must be checked which healthcare specific ethical principles, professional codes, laws and regulations, patient rights, and security and privacy rules are implemented, and how.
- How is the trustworthiness of the service provider measured?
- How are security and privacy managed in cross-jurisdictional healthcare services?
- To what extent do the service provider and the platform support openness and publish reasonable information for trust calculation?

Ruotsalainen et al. have mentioned that in e-health and in ubiquitous health services, providers often do not follow the security and privacy rules they have published (Ruotsalainen et al., 2013). Therefore, Ruotsalainen proposed that in e-health an external monitoring service that collects and analyzes trust and privacy features of service providers is needed. Based on analysis, it can inform customers of meaningful security and privacy findings. This kind of service can be a good addition for any e-health ecosystem (Ruotsalainen et al., 2014).

CONCLUSIONS

The e-health multisided market is a new service concept for the distribution of e-health services and PHI via information networks. It offers to patients, healthy persons, and potential sick individuals, a possibility to simultaneously use digital healthcare and nonregulated health services. Services of the e-health multisided markets are mediated and orchestrated dynamically by a common computer platform. Service providers and customers, as well as the multisided platform itself, can locate anywhere in the information space, i.e., in different jurisdictions. Customers and service providers are connected to the platform via commercial networks such as the Internet and mobile networks. This all means that the multisided e-health market and its services take place in an insecure information space. Therefore, security, privacy, and trust challenges existing in the multisided markets are very similar to concerns that exist in ubiquitous health systems (Ruotsalainen et al., 2013). For example, in ubiquitous health and in e-health ecosystems predefined trust cannot be automatically expected. A meaningful problem is also that in healthcare information systems, currently widely used security services such authentication and role-based access control and consent are ineffective in the environment of the e-health ecosystem.

In e-health and digital health large amounts of PHI and metadata are collected more or less ubiquitously. This makes it difficult (in many cases even impossible) for the customer to manage his or her information privacy, use the right to information autonomy, and set policies (rules) for controlling how his or her PHI is processed and shared inside the e-health ecosystem (Ruotsalainen et al., 2015).

It is evident that ethical challenges when e-health and digital health services are used via the multisided platform cannot be answered by expecting that all stakeholders apply principles of medical ethics. In the Internet and social networks it is typical that in spite of a service provider promising to follow principles of information ethics, and to be compliant with regulations of the customer's home country, there are few or no possibilities for verifying this. Instead, in many cases service providers have their own business ethics and business benefit focused privacy rules. To make e-health multi-sided markets (and the whole e-health ecosystem) trusted, a holistic framework that is built on common ethical principles, an advance-made trust verification and contextual privacy model is needed.

A challenging regulatory question is which regulatory model is most suitable for e-health and digital health. One solution is to move the regulatory privacy model used in current healthcare to e-health and digital health markets. Another possibility is to use the self-regulation model that is widely used in e-commerce (i.e., the service providers define rules and develop standards). Currently, healthcare specific laws and standards regulate only services provided by healthcare professionals. The regulatory model is also static and laws in different countries vary. It seems unrealistic to expand the patient status for all customers of the multisided e-health markets and to adapt the regulatory privacy model. In e-commerce, there is asymmetry between customers and service providers, and the customer is always the weaker partner (Ruotsalainen et al., 2015; Howell, 2006). In e-health and digital health, where the customer is an active and independent actor who makes personal decision, the self-regulation model is unacceptable. Ruotsalainen et al. have proposed that in e-health and ubiquitous health a model where privacy is the customer's personal property should be used (Ruotsalainen et al., 2013).

Another challenging question is the ownership of the PHI. PHI itself is much wider concept than "protected healthcare information (i.e., the content of the EHR)". Instead, PHI is "information about an individual person that relates to the physical and mental health of an individual" (ISO 27799 Health informatics—Information security management in health using ISO/IEC 27002), i.e., it includes the content of the EHR and the PHR (International Standardization Organization, ISO 27799, 2008). In healthcare the ownership of the patient information (and the EHR) is not in most of countries clearly defined. Typically, the care provider or the healthcare organization has by the law and regulations the responsibility to securely maintain and store patient records. The patient has limited rights to use opt-in or opt-out, and in this way controls the use of the content of his or her EHR (e.g., the distribution of the EHR). Concerning the ownership of the personal health record (PHR) no global consensus exists. In most cases, the person owns the content of the PHR, i.e. the individual, or his authorized representative, manages and controls the content of the PHR, and for example allows or denies clinicians participating the care to access the content of the PHR (International Standardization Organization, ISO TR 14292, 2012). In e-health multisided markets both healthcare and health information is communicated and processed. From individual's autonomy, independence, and privacy point of view it is fundamental that he or she owns the content of the PHR and regulates its use now and in the future. On the other hand, principles and rules regulating the management and use of the EHR are widely accepted and defined in healthcare specific laws and norms. This all means that the e-health ecosystem should have the ability to support both static healthcare specific and dynamic customer defined security and privacy rules. The new understanding of disease requires the availability of information which typically is stored in the PHR (e.g., environmental data, social data, epigenetic data, and personal health data collected before any sickness signs can be recognized). This has raised the need for easy ways to share the content of the PHR with clinicians. There are many other groups seeking the possibility to use the content of the PHR. For example, researchers, public health planners, and the pharmaceutical industry have increasing interest in using PHI. This means that in e-health ecosystems the PHI is used in different contexts. This is a strong argument for the use of a contextual privacy model and for supporting personal policies in e-health ecosystems.

In a multisided e-health platform the patient that uses healthcare services is not anymore a traditional passive object. He or she can be an active e-health service user, an informed chooser of services and increasingly the manger and coordinator of his or her own sickness and health (Ruotsalainen et al., 2015). This indicates that the patient role in e-health must be redefined, and the platform should have services that enable the patient to select dynamically roles he or she wants.

There are challenges still left in the way of trustworthy e-health ecosystems. The technology is not the barrier because we already have necessary ICT technology, guidelines, and standards (Seppälä et al.). The main challenges are political, emotional, and organizational. First, is not clear if healthcare professionals and existing healthcare organizations have a willingness to accept a paradigm change to multisided e-health markets in the form of competition between services providers and customer freedom to select services. Instead, many of them may want to gain the situation whereby all e-health services are part of the current one-sided healthcare information system. It is also possible that the e-health industry develops its own e-health markets based on self-regulation and industrial business rules.

The trustworthy multisided e-health markets require a comprehensive regulatory and legal framework. Security and privacy rules cannot be left alone in the hands of service providers and the e-health industry. Trustworthy multisided e-health markets need an updated legal framework that strengthens the customer's autonomy and status in such a way that he or she can validate in advance the trustworthiness of both the platform and all the service providers. The customer should also have power to define their own context-aware privacy policies based on the service provider's trust value or rank. It is also necessary to develop criteria for evaluation and certification the functionality, security, privacy, and trustworthiness of all stakeholders participating in the multisided e-health markets.

REFERENCES

Beyeler, W., Kelic, A., Finley, P., Aamir, M., Outkin, A., Conrad, S., Mitchell, M., Vargas, V., Creating interaction environments: defining a two-sided market model of the development and dominance of platforms, Complex Adaptive Systems of Systems (CASoS) Engineering, Sandia National Laboratories, Albuquerque New Mexico, USA {webeyel, akelic, pdfinle, msaamir, avoutki, shconra, micmitc, vnvarga} @sandia.gov.

Blobel, B., 2002. Analysis, Design and Implementation of Secure and Interoperable Distributed Health Information Systems, Studies in Health Technology and Informatics, Vol. 89. IOS Press, Amsterdam, ISSN 0926-9630.

Boudreau, K.J., Hagiu, A., Where platforms rules: multi-sided platforms as regulators, M2012 - GAWER PRINT.indd 164.

Cavoukian A., Privacy by Design, The 7 Foundational Principles. <https://www.ipc.on.ca/english/Privacy/Introduction-to-PbD/>.

de Reuver M., Keijzer-Broers W., Trade-offs in designing ICT platforms for independent living services, Delft University of Technology, Delft the Netherlands. g.a.dereuver@tudelft.nl

Digital healthcare: the impact of information and communication technologies on health and healthcare, the Royal Society, 2006, London, England, ISBN-13: 978-0-85403-636-3.

Eisenmann, T., Parker, G., Van Alstyne, M.W., Strategies for two-sided markets, Harvard Business Review On-Point Article, Product 1463, Boston University School of Management. <www.hbr.org>.

Evans, D.S., 2003. Some empirical aspects of multi-sided platform industries. Rev. Netw. Econ. 2 (3)

Gambetta, D., Trust: Making and Breaking Cooperative Relations. New York, NY: B Blackwell; 1988. Can we trust. <http://www.loa.istc.cnr.it/mostro/files/gambetta-conclusion_on_trust.pdf>.

Gawer A., Platforms, markets and innovation, M2012 - GAWER print.indd 43.

Heitkoetter, H., Hildebrand, K., Usener, C., 2012. Mobile platforms as two-sided markets, (Julyn 29). AMCIS Proceedings, Paper 11 <http://aiseI.aisnet.org/amcis2012/proceeding/AdaptionDiffusionIT/11>.

Howell, B., 2006. Unveiling 'Invisible Hands': two-sided platforms in healthcare markets, New Zealand Institute for the Study of Competition and Regulation Inc. and Victoria Management School, Article in SSRN electronic journal JUNE <http://dx.doi.org/10.2139/ssrn.913666>.

International Medical Informatics Association (IMIA). The IMIA Code of Ethics for Health Information Professionals. <http://www.imia-edinfo.org/new2/pubdocs/Ethics_Eng.pdf>.

International Organization for Standardization (ISO). 2008. ISO/IEC 27799 Health informatics—information security management in health using ISO/IEC 27002. Geneva, Switzerland. <http://www.iso.org/iso/home/store/catalogue_tc/catalogue_detail.htm?csnumber=62777>.

International Organization for Standardization (ISO). 2016. ISO/IEC FDIS 27011:2016(E), Information technology — Security techniques — Code of practice for Information security controls based on ISO/IEC 27002 for telecommunications organizations, Geneva, Switzerland.

International Telecommunication Union ITU, 2014. M2M enabled ecosystems: e-health, D0.2 – Version 1.0.

ISO PHR International Standardisation Organisation, ISO TR 14292, 2012. Health Informatics Personal Health Records — Definition, Scope and Context, <http://www.iso.org/iso/iso_catalogue/catalogue_tc/catalogue_detail.htm?csnumber=54568>.

Kostkova, P., 2015. Grand challenges in digital health. Public Health, 3: 134. <http://dx.doi.org/10.3389/fpubh.2015.00134>.

Lederer, S., Deay, A.K., Mankoff, J., 2002. UC Berkeley College of Engineering Technical Reports. Computer Science Division, University of California, Berkeley, CA, Jun. A conceptual model and metaphor of everyday privacy in ubiquitous computing environments, http://www.eecs.berkeley.edu/Pubs/TechRpts/2002/CSD-02-1188.pdf.

Margulis, S.T., 2003. Privacy as a social issue and behavioral concept. J. Soc. Issu. 59 (2), 243–261.

Mettler, T., Eurich, M., 2012. What is the business model behind e-health? a pattern-based approach to sustainable profit. ECIS 2012 Proceedings, Paper 61. <http://aisel.aisnet.org/ecis2012/61>.

Nissenbaum, H., 2009. Privacy in Context: Technology, Policy, and the Integrity of Social Life. Stanford University Press, Stanford, CA, USA©, ISBN:0804752370 9780804752374.

Noor, T.H., Sheng, Q.Z., 2011. Trust as a service: a framework for trust management in cloud environments. In: Bouguettaya, A., Hauswirth, M., Liu, L. (Eds.), Proceeding of WISE 2011, LNCS 6997. Springer-Verlag, Berlin Heidelberg, pp. 314–321.

Petronio, S., Jennifer Reierson, J., 2009. Regulating the privacy of confidentiality. In: Afifi, T.A., Afifi, W.A. (Eds.), Uncertainty, Information Management, and Disclosure Decisions: Theories and Applications. Routledge, New York, NY, pp. 365–383.

Rochet, J.-C., Tirole, J., 2004. Two-sided markets: an overview, <http://web.mit.edu/14.271/www/rochet_tirole.pdf>.

Ruotsalainen, P., Blobel, B., 2015. The new role of patient in future health settings. In: Blobel, B. (Ed.), Proceedings of pHealth 2015. IOS Press, Amsterdam. http://dx.doi.org/10.3222/978-1-61499-516-6-71. ISBM 978-I-61499-515-9.

Ruotsalainen, P., Nykänen, P., Seppälä, A., Blobel, Trust – base information system architecture for personal wellness, in Proceedings of the MIE 2014, Christian Lovis, Brigitte Séroussi, Arie Hasman, Louise Pape-Haugaard, Osman Saka, Stig Kjær Andersen (editors). IOS Press, Studies in Health Technology and Informatics Volume 205, 2014, ISBN 978-1-61499-431-2 (print) | 978-1-61499-432-9 (online).

Ruotsalainen, P.S., Blobel, B., Seppälä, A., Nykänen, P., 2013. Trust information-based privacy architecture for ubiquitous health. JMIR Mhealth Uhealth 1 (2), e23. http://dx.doi.org/10.2196/mhealth.2731.

Ruotsalainen, P.S., Blobel, B.G., Seppälä, A.V., Sorvari, H.O., Nykänen, P., 2012. A conceptual framework and principles for trusted pervasive health. J. Med. Internet Res. 14 (2), e52. http://dx.doi.org/10.2196/jmir.1972.

Seppälä A., Nykänen P., Ruotsalainen P., Development of Personal Wellness Information Model for Pervasive Healthcare, Hindawi Publishing Corporation Journal of Computer Networks and Communications. Volume 2012, Article ID 596749, 10 pages, <http://dx.doi.org/10.1155/2012/596749>.

Westin, A.F., 2003. Social and political dimensions of privacy. J. Soc. Issu. 59 (2), 431–453.

Telecommunication standardization sector of ITU, M2M service layer: Requirements and architectural framework, ITU-T Focus Group on M2M Service Layer, D2.1 – Version 1.0, 2014.

World Health Organisation <http://www.who.int/trade/glossary/story021/en/>.

PART IV

Implementation and Introduction of Technologies

CHAPTER 6

Sustainable and Viable Introduction of Tele-Technologies in Healthcare: A Partial Two-Sided Market Approach

L. Botin, P. Bertelsen and C. Nøhr
Aalborg University, Aalborg, Denmark

INTRODUCTION

It is a general trend in the Western world to develop telecommunication technologies with the aim of improving health services. A national mapping has shown that 273 projects exist in Denmark (Nøhr et al., 2015). However, only a few of them have been evaluated systematically and explicitly. A statement from the Capital Region (Copenhagen) showed that only 30% of the projects that were launched in the metropolitan area were evaluated (Region Hovedstaden, 2013). Among the ones evaluated, it is difficult to point to significant tangible health outcomes. In this chapter we will present concepts, values, and methods that can clarify, assess, and treat the development and deployment of telecommunications technology in healthcare systems. Throughout the chapter we have consciously chosen to use tele-technology in healthcare as a neutral overarching term to cover the various concepts (telemedicine, telehealth, telecare, teledialog, etc.), which on an everyday basis are used in technology-mediated health communication and treatment.

The aim of this chapter is to relate our enquiries and analyses to the overall "two-sided market" approach of this volume, where we find that there are two users/clients/consumers, i.e., healthcare systems and citizens/patients.

In order to embrace both the healthcare system and the real life-world of a very diversified population of citizens and patients, we think that methods have to be chosen from the outset with pragmatic principles and an emphasis on values and qualitative criteria to assess sustainable and viable solutions. This also means that the methods differ in focus in relation to the introduction of telecommunication technology solutions.

- "Clinical simulation" illustrates how the healthcare professionals interact with telecommunication technologies and how communication takes place and can take place between different health professional actors and the patient/citizen.
- The values represented by "value sensitive design" illustrate empathic approaches to understanding the users' needs, wishes, and requirements, where these are not

clearly explicated. Citizens, patients, families, healthcare professionals, etc. are not always aware of the values and norms a technology must pursue in an effort to support a better life.

- "Participation in networks" points to solutions where socially vulnerable and/or marginalized citizens are able to contribute to the design of technology with a potential for support and prevention. This ability will be enhanced through participation in networks.

The hypothesis is that the concepts, values, and methods have different focuses. "Clinical simulation" emphasizes how to involve the healthcare professional and the citizen while supporting a high degree of knowledge sharing across sectors. "Value sensitive design" primarily focuses on citizens, patients, families, and informal carers and how their quality of life is increased by interaction with telecommunication technologies. "Participation in the network" has a comprehensive socio-technical perspective on the issues where the purpose is to challenge the telecommunication technological solution in a social context where users are either unfamiliar with technology (e.g., the elderly), or unfamiliar with using the health-promoting functions (socially vulnerable or marginalized), and/or citizens who are otherwise marginalized in relation to the actual technology.

The choice of values and methods serve to improve the introduction of telecommunication technology in healthcare. This basically includes empowering, emancipating, and enhancing the conditions of life of the individual actor and/or group. Additionally it serves to ensure equal, fair, and democratic telecommunication technology solutions, where values and standards in support of this are visible and explicated.

In the following paragraph, we will discuss the confusion in terminology that exists around the concept of telecommunication technologies. We will further describe how it implies that professions and citizens have difficulty understanding each other. This potentially blocks the development and design of sustainable and viable technologies that create value and makes sense for those involved inappropriate two-sided market technological solutions. This terminology confusion often prevents equal, fair, and democratic system solutions for the good of the citizen, the health professionals, and healthcare as a social institution.

TERMINOLOGY

Evaluating the English Whole System Demonstrator Project (2009–12), English health psychologist, and expert in health informatics Stanton Newman operates with three main definitions of health informatics performed at a distance (tele):

- Telecare
- Telehealth
- Telemedicine
  (Department of Health, 2011).

Telecare

Telecare is available in several generations and is characterized by relatively mechanical, automated, and passive processes. Here technology detects the patient and the spatial context through alarms, sensors, etc. (Newman, 2014). An example of Telecare is found in the "Future Nursing Home" in Aalborg (DK) (Aalborg, 2014). Here the residents' small apartments are equipped with sensors in the floor that make it possible to detect movements and thus detect if a resident falls. This concept of telecare is, compared to our interpretation, problematic because the citizen is monitored and surveyed in an instrumental and mechanical way. This can, at first impression, be experienced as a distorted interpretation of the concept of "care," as it should mean to "care for," i.e., one of the core values of the healthcare professionals' work. This value is transformed in the technology to instrumental monitoring and surveillance, which is far from the caring and empathic approach that the value in its origins represent.

The Dutch STS researcher Jeanette Pols discusses in the book "Care at a Distance—on the Closeness of Technology" (Pols, 2012) the relations and issues connected to "warm hands" and "cold technology." Pols illustrates, through her ethnographic studies of chronic patients' use and relation to telecare (and telehealth and telemedicine) technologies, how this setup is wrong. Patients do not necessarily perceive technologies as cold and hands as warm. What counts is the good relationship between caregivers and care recipients, and how the good relationship can by supported by and mediated through technology (Pols, 2012, pp. 25–28). Telecare technologies are often invisible to the patient, because they are located in the walls, floors, curtains, etc. Thus, they can by first impression seem "cold." But as with other invisible background technologies, they perform the essential tasks of maintaining a life with quality. Both temperature and light in our homes are regulated by technologies that are invisible to us in our everyday lives, although that does not mean that we perceive them as alienating or cold. Quite the contrary, technology can be warm because it is discreetly watching us and reacting if something happens to us, or if we express a need, and in contrast to this, hands can be cold due to haste, lack of empathy, or personal animosity.

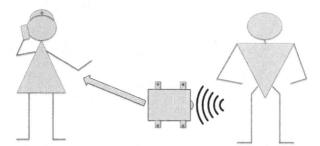

Figure 6.1 Telecare technology with mechanical, automated, and passive processes, technologies detect the patient and the spatial context and transmit signals to healthcare professionals.

Telehealth

Telehealth is characterized by some degree of activity from the user in relation to the technology. This means that the citizen, the patient, the caregiver, or the informal carer monitors conditions in the body and sends this data to a receiver in the healthcare system. It is worth noting that this receiver can be a server, and not necessarily a healthcare professional. This may have important implications for the behavior of the user in relation to the technology.

Citizens being active in relation to their health, by interacting actively with technology, characterize telehealth solutions. This means that the citizen must be able to carry out the activity, as well as being prepared to interact with his health and the available technology. It requires a physical and psychological readiness that not all possess. Therefore, the potential of the solutions should be carefully considered, in relation to equity, justice, and democracy. We will discuss this further in the context of the core values we have identified. Telehealth technologies are often more invasive than telecare technologies. Likewise, their presence is reflected in the rooms and/or in/on the body. They can also be invasive in terms of social relationships where the disease has become chronic, because a constant self-monitoring is needed. The technology becomes salient and mediates contacts and content between individuals. Brodersen and Lindegaard (2015) is an example of how the relative of a diabetes patient experienced and how their lives had changed to exclusively deal with the spouse's illness. This despite the fact that the specific medical condition was not serious (Brodersen and Lindegaard, 2015, p. 82). Another example, based on our knowledge base, is a remark from an elderly woman who, after her husband had been hooked up to a heart monitoring system, had ceased to have sex. She was convinced that there was a healthcare professional watching the man's heart 24/7, and she would be ashamed if sexual activity would be noticed and/or observed, or if signals otherwise would be misunderstood by the recipient.

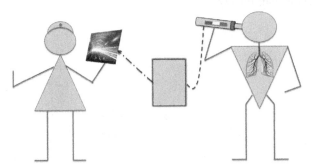

Figure 6.2 Telehealth technologies where the citizen, the patient, or the caregiver has to be active to monitor conditions in the body, which are then made available to healthcare professionals.

Telemedicine

Telemedicine, in "Whole Systems Demonstrator" and Newman's definition of the term, is characterized as supporting communication between healthcare professionals. It means that the patient, citizen, etc. are target fields for diagnosis and decision-making within the sectorial framework of the institution. Examples of this include homecare providers who send a picture of a wound to a doctor to obtain direct indications regarding wound care. Another example could be the transfer of information/communication from one sector to another via a telemedicine technology. Here too, it often involves talking about pictures, scans, and/or test results. This same data can also be transferred in the sector between different groups of healthcare professionals. Telemedicine thus indicates very concrete exchange of information through telecommunication technology among healthcare professionals. It can be performed in the citizen/patient's home, but as stated it is the healthcare professionals who are using the technology, and citizens/patients are physical or virtual objects. It is important when developing telemedicine solutions to ask the question to what extent—and how—the measuring object should be involved in the process. How do we create solutions that secure equity, justice, and democracy within the institution? How is it done so that it will not be specific occupational groups who will benefit from the technology at the expense of others being more knowledgeable and/or competent in the use of technology and understanding, i.e., viable, responsible, and sustainable two-sided market solutions.

THE MYSTERIES OF HEALTH

The German philosopher Hans-Georg Gadamer pointed out that, as long as our health is invisible to us, we will experience our health as functioning (Gadamer, 1996). Therefore, it can be seen as a paradox that we develop tele-technologies that constantly put our health in focus and makes it visible to us in our everyday lives. This is not a problem from

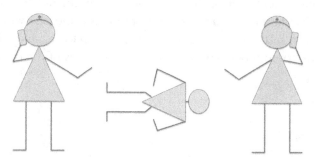

Figure 6.3 Telemedicine technologies support communication between healthcare professionals and patients or citizens.

the moment we have lost our health and have become ill, i.e., for patients. However in a prehospital context, where we monitor our body through "quantified self" technologies, it does become a problem, from Gadamer's perspective. The problem is that these "persuasive" technologies are constantly visible and exhibit the physical condition of our bodies. The psychological and experiential consequences of this technological visibility of our bodies can lead to exaggerated and inappropriate behavior that ultimately destroys our health. Constant self-monitoring has the potential for such inappropriate behavior where the technical optimization, on the basis of figures, becomes the target of the way we govern ourselves in our everyday lives (self-management).

As indicated in the introduction, we intend through our partial two-sided market perspective to empower, emancipate, and enhance the healthcare professional, the citizen, the patient, the relatives, and informal carers in relation to the use of telecommunication technology solutions. This is our conceptual take on the two-sided market discourse, where we look at the definitions of Evans and Scmalenmsee (2013) "a multisided market has two or more groups of consumers, who need each other, who cannot capture the value of their mutual attraction and who rely on a catalyst to facilitate their interaction" (Evans and Scmalenmsee, 2013). From this perspective telecommunication technologies are seen as catalysts for appropriate interaction in between health professionals and citizens/patients; two sides that need to be bridged in order to create a viable and sustainable ecosystem.

For that reason, it is not the case that some must suffer losses of power, position, etc. at the expense of others in this process. On the contrary, the focus is on securing and supporting all stakeholders' visibility activities, understanding, and application of technologies for creating the framework for better health and care (prevention, treatment, and rehabilitation), for less money. This means that stakeholders and actors representing the "systemworld" and "life-world," from a Habermasian perspective, are taken into account in order to reach communication and interaction without domination (Habermas, 1968, 2005).

In relation to this it is important to reach a common understanding of what is meant by "empowerment," "emancipation" and "enhancement." It is important since there is a similar confusion in relation to these terms as is the case with the aforementioned confusion in relation to e-health, telemedicine, telecare, and telehealth. In the following, we will try to provide some unique formulations of the content and meaning of the concepts.

EMPOWERMENT

The concept of empowerment has several different meanings, depending on what is to be illuminated by the concept. Fumagalli et al. (2014) indicate that most interpretations of the concept contain meanings as skills, motivation, and the ability to hold and exercise power. At the same time, however, they note that it is difficult to distinguish between empowerment from a patient's perspective and "neighbor concepts" such as involvement, participation, activation, engagement, and enabling the patient to perform actions. The difficulty in making this distinction creates confusion as well as danger of the concept being diluted, as we have seen in trends over the last few years. Fumagalli et al also point out that: "As a result we have a lot of valuable evidence that remains dispersed because different research streams struggle to communicate" (Fumagalli et al., 2014, p. 385).

These flaws or deficiencies in communication should be attributed to paradigmatic affiliation, wherein different paradigms/disciplines ascribe the concepts different meaning, or have the same beliefs though using different terms. Fumagalli et al. compiled a formulation of empowerment, whereby they seek to explain the "neighbor concepts" in relation to the overall and wider concept of empowerment, is as follows:

Patient empowerment is the acquisition of motivation (self-awareness and attitude through engagement) and ability (skills and knowledge through enablement) that patients might use to be involved or participate in decision-making, thus creating an opportunity for higher levels of power in their relationship with professionals.

Fumagalli et al. (2014, p. 390)

Motivation and skills through engagement and enablement are prerequisites for empowerment and activation. Activation is specifically geared to the specific illness, and empowerment has a more holistic perspective on self-esteem and quality of life. Concerning behavior, it is controlled by the involvement and participation, where the distinction is somewhat unclear in comparison (Fig. 6.1). Involvement influences empowerment, and empowerment is a prerequisite for participation. This means that someone must surrender power to the patient so that the latter is able to participate. Fumagalli et al., is of the persuasion that building skills and boosting motivation are the main aspects related to patient empowerment. In contrast, they are more critical toward the notion of possessing and exercising power, as the latter do not necessarily need the desired outcome. A patient can exercise his power in contradiction with the expert knowledge of a healthcare professional and act inappropriately in relation to their own health. Therefore, the involvement and participation must be placed into critical/reflective environments in which the involvement and participation can be evaluated in relation to the desired outcome. In other words ask the question—does the involvement and participation of the patient enhance the patient's health and quality of life and does it protect the patient as an individual and a citizen? (Fig. 6.2), (Fig. 6.3), (Fig. 6.4).

"The patient at the center" has been a mantra in hospital services for almost a decade, and much has been made to support this position. Telecommunication technological solutions are written into this paradigm and context where the concept of empowerment is central. It is clear that the concept of empowerment is not only limited to the patient, but actually should be broadened to cover all players related to telecommunication technologies. Healthcare professionals ought also to experience being

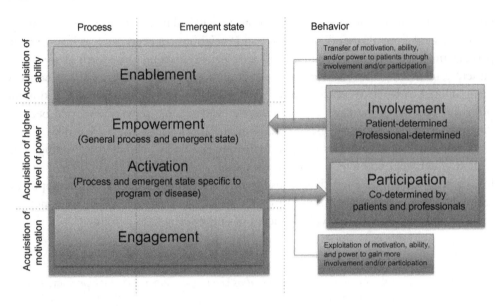

Figure 6.4 The concept of patient empowerment related to involvement and participation (Fumagalli et al, 2014).

empowered in their daily activities, as should relatives and informal carers. In order to be empowered in their work, they should be engaged, educated, activated, involved, and participatory in the development and design of telecommunication technological solutions. The consequence, if this does not happen, will be that healthcare professionals, relatives, and informal carers will be "invisible" in our focus on the patient, and they will experience a loss, and disempowerment. This will consequently imprint on their behavior in relation to telecommunication technological solutions in healthcare.

Table 6.1 is structured on the basis of Alpay, Van der Boog, and Dumaij's study of how we create "innovative e-health tools for self-management." It identifies some issues related to healthcare professionals' competencies and skills when it comes to empowerment of the patient, where there is a focus on how the healthcare professional experience is being empowered in a new and dynamic reality in which roles, significance, and meaning are constantly changing (Alpay et al., 2011).

Overall, it is distinctive regarding its dynamic and procedural concept formation, as the reality that characterizes the relationship between actors and stakeholders is constantly changing. In some cases it is necessary that the healthcare professional have the opportunity to take on a maternalistic/paternalistic and controlling role in the communication and interaction with citizens, patients, relatives, and informal carers, since they do not have sufficient knowledge and experience and/or are too weak, and physically or mentally unable to take care of their own health and rehabilitation.

Table 6.1 Skills to be supported by healthcare professionals when implementing telemedicine with the goal of increasing patient empowerment

Component to the development of empowerment	Skills possessed by the healthcare professionals
Communication	Skills in using different communication channels targeted at patient's needs
	Access to information from other professionals
Education and health competencies	Imparting patient knowledge
	Knowledge of the information and training available, convenient to the patient's needs
	Insight into the patient's knowledge and skills
Information about the patient's state of health	Insight into the patient's knowledge and skills
	Being available to inform the patient
Self-care	Entrust the control of taking care of health to the patient itself
	Be willing to accept the patient's health requirements and preferences (switching between recommended care and patient-centered care)
	Be prepared to meet the patient's wishes, opportunities, and situation when making decisions
	Supporting patients
	Coaching patients
	Skills in having motivational talks with patient
Decision support	Sharing clinical experience/knowledge with the patient
	Knowing the patient's ability to make decisions
	Knowing the patient's choices/options
Contact with fellow patients	Being aware of opportunities for patient networks
	Being aware of the benefits and risks for patients

The table is based on Alpay et al., 2011 and Nielsen et al., 2014.

There is thus a risk that other stakeholders and actors are invisible and become disempowered in relation to the "strong patient" and further that the "weak patient" will not be seen, in the attempt to strengthen the already strong.

To ensure salience and prevent loss and disempowerment, we believe that a new concept in relation to the introduction of telecommunications technology in healthcare can be introduced. In the following we will explain how emancipation can complement empowerment in a fruitful and appropriate manner.

EMANCIPATION

The concept of empowerment is well known and frequently used in the literature, whereas the concept of emancipation is almost absent in relation to healthcare information technologies. We believe, from a value and attitudinal perspective, that the problems occurring in connection with the implementation of telecommunication technology in healthcare can be addressed with great advantage from an emancipatory angle. The emancipatory perspective is likewise the point of departure for selecting methods and how we describe, analyze, and evaluate these methods.

The German sociologist and philosopher Jürgen Habermas describes emancipation as dialectical in its essence. This means that the moment we examine/design technological solutions, there is an ongoing emancipation of both the examined and the examiner. The relationship between the examined and the examiner is likewise dialectical. It opens up to emancipate the investigated in the process. This means that the focus of study is activated in an interaction which ensures that emancipation takes place within a framework that supports both "system-world" and "life-world." Habermas writes:

> The methodological framework that determines the meaning of the validity of critical propositions of this category is established by the concept of **self-reflection**. The latter releases the subject from dependence on hypostatized powers. Self-reflection is determined by an emancipatory cognitive interest. Critically orientated sciences share this interest with philosophy.
>
> **Habermas (1971, p. 310)**

The healthcare system holds a hypostatic power, manifested in the hierarchy and accompanying procedures of the institution that are also found in the ways technologies are developed and applied. Having no self-reflective and emancipatory perspective on technology development, the hypostatic power will manifest itself in technologies. Thus it will automatically be carrier of an instrumental and technical rationality that prevents expression of the subject's/individual's life-world. The purpose of the emancipatory cognitive interest is to open the space for a rational and supremacy-free dialog where all contributors to the dialog will be "released" in the process. This means that not only is the patient emancipated, but so are the healthcare professionals, relatives, and informal carers. They will find that their democratic autonomy is supported and respected, in other words the concept of citizenship is highlighted and qualified.

In his essay "Science and Technology as Ideology" (1968), Habermas presented a table (Table 6.2) for understanding the relationship between a life-world's rationality and a system-world's rationality.

Choosing to read the table dialectically, the two "worlds" talk to and touch each other and this intertwinement provokes change in the actual conversation and touch. Instrumental and strategic action cannot stand alone, because the result would be hypostatized exercise of power mediated through technology. The life-world perspective cannot stand alone either, because it is driven by established black-boxed norms and habits, and thus basically reactionary and conservative in its regulations and functions. We should aim for a dialectical and critical bridge between technical rationality and communicative rationality, because in this way we ensure that the health professional experts' knowledge and practice come to fruition while we give room, space, and time to the actors' life-world to flourish within context-sensitive frameworks.

Table 6.2 The relationship between a life-world rationality and a system-world's rationality

	Institutional framework: symbolic interaction	Systems of purposive-rational (instrumental and strategic action)
Action-orienting rule	Social norms	Technical rules
Level of definition	Intersubjectively shared ordinary language	Context-free language
Type of definition	Reciprocal expectations about behavior	Conditional predictions Conditional imperatives
Mechanisms of acquisition	Role internalization	Learning of skills and qualifications
Function of action type	Maintenance of institutions (conformity to norms on the basis of reciprocal enforcement)	Problem-solving (goal attainment, defined in means-end relations)
Sanctions against violation of rules	Punishment on the basis of conventional sanctions: failure against authority	Inefficacy: failure in reality
"Rationalization"	Emancipation, individuation; extension of communication free of domination	Growth of productive forces; extension of power of technical control

Habermas, J. (1968).

Developers should consider this two-sided and dialectical approach as they design together with users (healthcare professionals and patients/citizens), where they as the third party bridge the interests, requirements, needs, and wishes of the "market"

ENHANCEMENT

Technology has always been intended to enhance human capacities and potentials. Telecommunication technologies ranging from smoke signals to telemedicine have been developed over time with the aim of spreading information over distance, and technology development has led to this availability being independent of space, time, and place (Commission, 2012). Telecommunication technologies are emblematic and exemplary for technology's technical potential, and by its intervention we can probably gain knowledge about everything. Telecommunication technologies are enhancement, and it is obvious that the forces to be found in technology as a technique in some way have to be managed and given a direction. This means that the purpose of development and design of telecommunication technology must and should be subject to considerations that discuss how technology enhances our understanding of telecommunication technology's potential and simultaneously amplify and enhance the processes and conditions that support values such as equality, justice, and democracy (Commission, 2012).

The concept of enhancement is important in relation to the previously identified different definitions of telemedicine, telehealth, and telecare in terms of the methods we present below. In which direction and for what purpose do we use data that we as citizens gather about ourselves? In what way is quality of life amplified and enhanced through this data collection? When aspects are enhanced and/or increased does this then happen for a specific purpose? For instance, are minorities or marginalized groups enhanced and strengthened through their use of telecommunication technology in order to achieve an opportunity for a better self-esteem, a better position in society, and thus also better health?

The American researcher and feminist Donna J. Haraway philosophically, socially, and politically contextualizes the information associated with the enhancement concept in technology. Whilst Habermas is of the basic assumption that technology must be tamed from an emancipation perspective, Haraway has a much more positive view of technology's potential in relation to the solution of power structures and ingrown prejudices about gender, race, and the like (Haraway, 1991).

Haraway suggests an opportunity in this relationship for deep integration and understanding where the distinction between humans and technology is impossible and to some extent also absurd. She writes:

> Intense pleasure in skill, machine skill, ceases to be a sin, but an aspect of embodiment. The machine is not an it to be animated, worshiped, and dominated. The machine is us, our processes, an aspect of our embodiment. We can be responsible for machines; they do not dominate or threaten us. We are responsible for boundaries; we are they.
>
> **Haraway (1991, p. 174)**

This is the positive image of the cyborg and technological enhancement, but there is also a dystopian one of its kind that we must keep in mind the moment we support the formation/creation of digital cyborgs (see below), and the image is carried by an instrumental, technical, and systemic understanding of reality that largely refers to Habermas'. The cyborg created in this vision looks as follows:

> Control strategies will be formulated in terms of rates, costs of constraints, degrees of freedom. Human beings, like any other component or subsystem, must be localized in a system architecture whose basic modes of operation are probabilistic, statistical. No objects, spaces, or bodies are sacred in themselves; any component can be interfaced with any other if the proper standard, the proper code, can be constructed for processing signals in a common language.
>
> **Haraway (1991, p. 167)**

Telecommunication technologies have a potential to develop "monsters" of a similar nature, if we do not methodologically address this potential with the aim of fostering their counter-image, namely existentialist and responsible telecommunications technology solutions.

Enhancement is done by "cross borders," so that new and unprecedented configuration manifests itself in representations that are not directly related to something already existing. This means that the cyborg is not necessarily an evolutionary size where humans have perfected/optimized themselves in interaction with technology. It can also be an absolute breakthrough (revolutionary), where there was no link to something original, and its behavior does not have to be logical or rational. It has an impact on how we view the potentials inherent in the development and design of telecommunication technology solutions, as the Australian sociologist Deborah Lupton argues in her application of Haraway's conceptual framework related to digital telecommunication technology (Lupton, 2013, pp. 1–15). Lupton uses the term "the digital cyborg assemblage" to define the conditions that apply when: "… we understand our bodies/ even through technologies and our bodies / even makes sense and configure the technologies through our co-actions in everyday life" (Lupton, 2013, p. 5). It is the way things are assembled and how they work together that constitute enhancement where our bodies are seen as "… a complex and dynamic configuration of their own bodies, the bodies of others, discourses, practices, ideas, and material objects" (Lupton, 2013, p. 6). Lupton's cyborg assemblage can act in two different ways respectively, to point to the evolutionary and revolutionary. From an evolutionary perspective it is viewed as generally conservative with a focus on ideals, as a whole, of purity and self-responsibility, which are largely supported by the healthcare system and its institutions/actors. But from a revolutionary (disruptive loss) perspective the cyborg assemblage opens up to the individual's well-being, where health is improved and strengthened: "… while maintaining a critical distance to how individuals and groups can be oppressed, stigmatized, and excluded by these technologies, and how the rhetoric and practice of digital health serves powerful interests" (Lupton, 2013, p. 12).

Enhancement has the same dialectical and critical potential as emancipation. Yet it emphasizes to a greater extent the interaction between technologies and humans.

CONCLUSIONS ON TERMINOLOGY AND CONCEPTS

In this section we have focused on the ambiguity that conceptually exists in the field of telecommunication technology related to health. There is a need to clarify concepts, and we have tried set the foundations of this conceptual clarification. We have done so on the basis of Stanton Newman definitions of telecare, telehealth, and telemedicine, although we are aware that these are also subject to a certain kind of weakness and that the boundaries can be difficult to draw. But overall, we can conclude that telecare is characterized by fixed and static technologies that collect data about the citizen/patient at home and send them on to the system. Telehealth is characterized by fixed and mobile technologies where the citizen/patient, relatives, or others, tap in measurements etc., and then through telecommunication technology send them into the healthcare system. Telemedicine is telecommunication technologies, which develop information, and facilitate communication between healthcare professionals in different sectors and with different disciplines. In this case, the citizen is the subject matter of information/communication. As noted, the boundaries between the different telecommunication technologies are continually blurred, and in some cases there are technologies that are both, or cannot be classified as either. Despite this, we believe that Newman's classification system can be used advantageously to overcome the apparent confusion that exists, and thus can prepare the ground for meaningful interaction between the various actors involved in the development and design of telecommunication technology in healthcare. We also believe that this is the condition for building health literacy because without paradigmatic concepts, in which we agree on how to understand the concepts, we cannot be competent and/or experts, but basically remain illiterate.

In the second part of this section, we discussed how the players can be activated in a meaningful way with the aim of creating a sustainable and viable introduction of telecommunication technology in healthcare. We have emphasized that all players should experience empowerment, emancipation, and enhancement. We believe that citizens/patients, caregivers, informal carers, and healthcare professionals must experience being strengthened in their life and work and that empowerment cannot be at the expense of other actors in the process. To support this, we have emphasized that empowerment should be complemented by emancipation and enhancement, because a self-reflective, dialectical, and critical angle means that things are really sustainable and viable. At the same time we are looking through the enhancement concept of attributing telecommunication technologies with a positive potential in relation to the care for our health and the resolution of the general picture of technology as something cold and basically inhuman. In the following, we will describe three qualitative methods that we recommend in relation to appropriate and effective approaches to development and deployment of telecommunications technology solutions in healthcare. In doing so we try to address, in relation, how "groups of consumers" can be dealt with and "mutually attracted" by mediating technology, where designers and developers delegate "catalyst" potential qualities into the technology, hence taking a partial two-sided market approach (see Chapter 1: An Ecosystem Perspective on Two-Sided Markets in e-health).

USABLE METHODS

Clinical simulation as a method to evaluate and assess the use of e-health in complex work processes.

A clinical simulation can enable an evaluation of the actual use of a prototype in a realistic environment. This approach is applicable to the assessment of potential consequences, cognitive processes, usability, and applicability. A more detailed and elaborate presentation of clinical simulations is described in Jensen et al. (2015).

Figure 6.5 Basic steps in a clinical simulation.

In general, a clinical simulation involves four steps: Purpose, planning, preparation, and performing (Fig. 6.5).

Purpose of the simulation

Clinical simulation can be a very beneficial method to use in the initial phases of a design project to explore work practice and develop user requirements. Later in the project the purpose can include implementation issues, appraise the need for education and training, and analyze the influence of new technology on existing work processes.

In the design phase clinical simulation can be applied as a "boundary object" to expose conflicts of interest and to gain consensus among different stakeholders. This could be very relevant for e-health products trying to reach a two-sided market. Clinical simulation enables different users to observe the practical use of new e-health systems in real time, and the subsequent debriefing and discussions provide an opportunity for a deeper understanding of new procedures and requirements. This can be seen as a participatory approach to design where future users are engaged actively in the design process and thereby influence design solutions directly (Rasmussen et al., 2012). Clinical simulations can also be initiated by one or more design workshops where different stakeholders build prototypes or provotypes (Boer et al., 2013). Furthermore, clinical simulations can assess the need for training and information prior to the implementation and "go live" of a specific system. Knowledge about work practice and patient safety issues can be achieved and utilized as important input prior to or during a pilot implementation process. As for any participatory project it is of paramount importance that the purpose of the simulation is stated and agreed upon from the very first stage (Kuwata et al., 2006). This should be done in close cooperation with the stakeholders and the owner of the project (Jensen et al., 2014).

Planning of the simulation

In the next phase the scope for the simulation should be defined and it should be determined which scenarios should be used in the simulation. It should also be decided how many rounds of evaluation are necessary in order to obtain a significant result. Furthermore, the profiles of the participants must be determined. The number of rounds of evaluation depend on the number of scenarios necessary for the evaluation, the number of participating clinicians and patients/citizens, and the purpose of the evaluation.

The choice of scenario is crucial and must reflect the purpose of the evaluation. Each scenario reproduces a typical and significant work task, and the sum of scenarios should cover the work practice that is affected by the new technology. Scenarios are constructed as narrative descriptions of people and their activities (Carroll, 2000). Scenarios can be constructed to focus on specific goals defined for the purpose and to highlight the appearance and functioning of the technology, and how people will interact and perform with it. In the scenarios, various environments and settings can be set up for the occasion and additionally supplied with sound to make them realistic to the actors who perform the script from the scenarios. The choice of scenarios will affect the entire evaluation and should be considered carefully to fulfill the purpose of the evaluation.

The profile of the participants in the simulation must be clearly defined. This concerns specifically the users representing patients or citizens. Clinicians are normally functioning in similar roles in their work life and adapt more easily to their given role in the simulation. However, it is usually easier to make experienced clinicians focus on the use of the technology as opposed to junior and inexperienced health professionals who tend to focus on their own performance. In cases where the purpose of the simulation particularly is focused on inexperienced use of technology this will clearly not be the rule. Actors representing patients or citizens are most critical and should be defined and described in more detail and selected with great care.

In cases of a simulation of a very complex work situation, involving many functionalities used by many different groups of healthcare professionals and diverse patients with complex multimorbidity diagnoses, the number of scenarios and simulation cycles must be greater. This will most often be the case when testing technologies that have both health professionals as well as patients as users.

Preparation of practicalities

Following the planning of the overall frame for the evaluation, the actual test should be prepared in detail. The scenarios have to be written, the actors to perform the simulation have to be identified and appointed, and the clinical and technical setup must be prepared (Lawtonet al., 2011). The clinical setup must reflect the real settings and if the simulation incorporates the patient's home this must be designed accordingly. The technology must also support the functionalities expected from the system to be designed and in accordance with the scenarios.

The resources needed to prepare a clinical simulation can be considerable and time consuming, depending on the complexity of the work procedures reflected in the scenarios. But also the fidelity of the whole setup demands resources and both aspects must therefore correspond diligently to the purpose defined in the initial phase (Ammenwerth et al., 2012; Dahl et al., 2010). It is however important to note that the time spent by the actors—doctors, nurses, secretaries, or patients—who perform the actual simulation is around three hours, including briefing, performance, and debriefing. The total planning, preparation, and running of a clinical simulation can take more than 100 h (Jensen et al., 2015).

Determining the adequate level of fidelity will be dependent on the purpose of the clinical simulations, and the phase of the lifecycle of the system evaluated, and there are a number of specific issues to consider: equipment fidelity, environment fidelity, task fidelity, and functional fidelity.

The equipment and functional fidelity corresponds to the phase of the system development lifecycle, and the task and environment fidelity is related to the purpose of the simulation study.

In cases where simulations are used to analyze user requirements they can be conducted with different degrees of fidelity. High fidelity tasks have well described scenarios and high fidelity equipment and functionality, in the form of mature prototypes with realistic test data, but also in a more experimental form with the use of low fidelity equipment and functionality (Jensen et al., 2013). A "wizard of Oz" methodology may be present in low fidelity simulations, where cardboard boxes replace equipment and a person simulates the response and functionalities from the system in form of handwritten post-it labels (Molin, 2004). The "Wizard of Oz" method offers interactive experience without having a real computer system. It can produce adequate and sufficient input to identify user requirements or explore key tasks in controlled environments.

Fig. 6.6 illustrates the stage in the system's lifecycle and the degree of fidelity and relationship to the purpose of the evaluation (Jensen et al., 2012).

Nothing is perfect in the first trial and it is worthwhile pilot testing the simulation and the scenarios before the participants are involved in the real evaluation. Separate pilots can be made on the scenarios, the clinical setup, the technical setup, the test data in the information system, and the data collection.

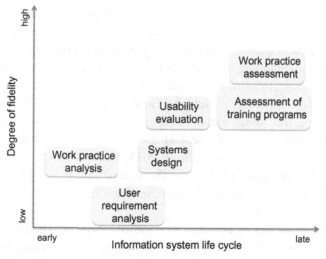

Figure 6.6 Clinical simulations in relation to information system's lifecycle and degree of fidelity.

Performing

Before the actual simulation embarks, it is important to introduce the participants to what is going to happen. This includes an introduction to the purpose of the simulation, and making it clear that it is the system that is being tested and not the participants' performance. The participants must also be introduced to the system and the functionalities which are utilized in the scenarios. The time spent on each task should reflect the purpose of the evaluation. In the case where intuitiveness is evaluated, users should not be allowed too much time to get themselves acquainted with operating the system, whereas in cases where optimizations of work procedures is in focus, it can be necessary to train the users comprehensively before they take part in the simulation. After the introduction to the system the users will use during the simulation, they are introduced to the environment and the scenarios, and this includes other actors who are part of the simulation and might appear during the performance of the scenario. Depending on the purpose of the simulation, various interrupting elements can be included in the scenario, and to ensure the maximum effect of these interruptions the participants should obviously not be informed about them (Ammenwerth et al., 2012). However, it is also important that the participants, clinicians as well as patients, feel comfortable in their role during the simulation so they can focus on the scenario and the technology rather than the simulation as such.

To help in conducting the scenarios a facilitator can be located in the simulation room—here she/he can support the clinician or patient in handling the technology and answering questions about details in the scenarios.

Depending on the purpose, the facilitator can either act as a "fly on the wall" and remain unobtrusive, or alternatively engage actively with the clinician. If a high degree of accuracy is required, the facilitator should be involved as little as possible in order to make the flow in the simulation natural. Any interruption will disrupt the clinician's cognitive processing and empathy in the simulation and reduce the perceived realism. If possible, ask the participant to "think aloud" during the simulation, for the observers to gain a deeper understanding of human behavior in the performance of the task. This method helps to uncover the more cognitive aspects of the interaction between users and technology, and is especially useful when analyzing users' needs. Depending on the purpose of the test, the "think aloud" method can be supplemented with observer-view, where a facilitator asks more comprehensive questions about the use of the system, and requirements thereto (Kragelund, 2013; Rasmussen et al., 2013). It is also possible to get the participant to describe the system and its functions in a relatively natural setting, by letting a "patient" or "colleague" ask questions about the system and its use during the simulation.

To create a high degree of accuracy in recreating the simulated environment, the participating clinicians should have great familiarity with the working practices which are being simulated. Often clinics and quality managers volunteer to be participating clinicians, but they are not familiar with the details of the actual work performed in the clinic. The same goes for the patient—the best result is achieved if it is real patients participating, or at least someone who has been a patient in the area that is simulated. Their knowledge of the participants must focus on how the work is actually carried out and not on how the work activities should be done. If participants are not familiar with the work practice, the simulations are being conducted under false pretenses, and the result will be of less value. Clinicians with extensive experience in testing and evaluation of health IT will often think of themselves as, and act as, testers instead of clinicians (Jensen et al., 2014).

Data gathering and analysis

Qualitative as well as quantitative methods are used in simulated evaluations.

After each set of simulation scenario is complete, the participating clinicians are asked to complete a questionnaire and a debriefing interview held with clinicians and observers. The questionnaires must reflect the purpose of the evaluation and can include questions or statements about efficiency and satisfaction, as well as questions or statements regarding simulations and realism of the scenarios.

The interview guide can be composed of open questions, starting with a few basic questions regarding the positive and negative features of the system. Subsequently, more specific questions are asked to let health professionals clarify and expand statements and questions from questionnaires and other specific issues as they come to mind. The composition of questions or statements should reflect the purpose of the test. Interviews may be held individually or in focus groups. At the end of each day, the data from the interviews are analyzed by the method "Instant Data Analysis" (IDA) (Kjeldskov et al., 2004). IDA is a cost-analysis technique that makes it possible to perform usability evaluations, as well as analyze and document them, in just a day.

In a case study from Aalborg University, it was shown that the IDA, in only 10% of the time required, performed a complete video analysis to identify 85% of the critical usability problems in the evaluated system. IDA is performed right after the evaluation has taken place, where observers and facilitators from the usability test participate. Based on observations and notes from simulations and debriefing interviews, usability issues are identified, described, and categorized.

Some simulation studies may require a more conventional analysis e.g., traditional video analysis or a "Grounded Theory" approach. These methods, however, can be resource-intensive, and choice of collection and analysis of data is recommended to reflect the purpose of the evaluation.

It should also be determined for whom the results will be presented and by whom the findings and recommendations will be used and implemented.

In summary, the ten most important points to pay attention to can be boiled down to:

- The purpose of the clinical simulation must be focused and anchored in the organization that will use the results.
- Choice of scenarios is crucial and must reflect the purpose of clinical simulation.
- Choice and profile of clinicians and patients or citizens must reflect the purpose of the clinical simulation.
- Complexity in scenarios and patient information must be carefully considered.
- Planning and preparing clinical simulation is resource demanding in order to make it time effective for clinicians and patients/citizens.
- Degree of fidelity must reflect the purpose of the clinical simulation and the maturity of the technology.
- Rehearsals and pilot studies are important and well worth the effort.
- Real clinicians and patients should be used as participants.
- Cost saving analysis methods like IDA are very useable and can be practically applied to analyze the resultant data.
- The mandate of the clinicians, the patients/citizens, and the observers, as well as how the results will be used, reported, and implemented, should be made clear to everybody.

Value sensitive design (VSD)

VSD was introduced in the 1990s as a response to modernism's emphasis on technology as a mechanical fix that could solve (all) existing problems. The basic assumption in the VSD is that technology-focused solutions often lead to "inhumane" technologies and practices in relation to technology. Methodically, VSD works with three elements, which are all in play at the same time. The three elements are: conceptual studies, empirical studies, and technical studies (Cummings, 2006). In the following we will sketch the basic characteristics of the method and subsequently investigate its potential in relation to the design, development, and deployment of telecommunication technology in healthcare.

Conceptual studies: this kind of study is based on general philosophical and ethical considerations. VSD delivers a list of concepts which technology developers can/ should take into consideration during the process: human dignity, justice, welfare, human rights, privacy, trust, informed consent, respect for intellectual property rights, universal/general usability, environmental sustainability, moral responsibility, general accountability, honesty, and democracy (Manders-Huits, 2011, p. 275). Additionally, it is necessary to investigate which stakeholders are directly and indirectly affected by the technology. This means that there is extensive stakeholder analysis, and stakeholder values are examined with the aim of ensuring that they are, as far as possible, taken into account in the process and the technology.

Empirical studies: this kind of study is based on established survey methods: interviews, focus group interviews, questionnaires, observations, and measurements of consumer behavior, etc. The studies are designed to determine if a technology conflicts with stakeholder interests and values, and overall the technology respects VSD's conceptual values.

Technical studies: in the study of a specific technology, emphasis is on identifying if the technology infringes the human and moral values that have been identified in the conceptual and empirical studies. This is a quite detailed study of the technology held up against the parameters that have been established in other studies.

In VSD it is important to note how all studies are in play at the same time and there seems to be a certain waterfall effect because especially the technical studies are the result of the previous studies. Likewise, the empirical studies are a result of the extended stakeholder analysis of the conceptual inquiries. This linearity is a bias in the method, because it can "freeze" the values fixed and block the recognition and reflection in the process. There are also other weaknesses in the method, as Noemi Manders-Huits points out in "What Values in Design? The Challenge of Incorporating Moral Values in Design" (Manders-Huits, 2011). Manders-Huits primary appeal points are: (1) the methodological ambiguity in relation to identifying stakeholders, (2) methodological ambiguity in relation to integrating empirical methods with conceptual studies, (3) the risk of making an *is* into a *should*, (4) value concepts and how they should come to fruition is undecided, and (5) VSD lacks a complementary or explicit ethical theory that can solve ethical dilemmas, i.e., create a hierarchy of values (Manders-Huits, 2011, p. 271).

Manders-Huits points to the need for a "value advocate" to ensure that there is an explicit ethical theory basis for priority ranking and selection of values. The values are relevant in relation to the concrete, and in that the stakeholders' values should be given priority, and in terms of which empirical methods should be used (Manders-Huits, 2011, p. 285).

The justified criticisms from Manders-Huits perspective have been taken up by another Dutch researcher Ibo van de Poel (2016), who has tried to translate parts of Manders-Huits criticisms in a model that can be used by multidisciplinary teams of design development.

Van de Poel works on a model that can be read/used top-down or bottom-up. At the top of the model a value or set of values is installed, then followed by norms to ensure the realization of the value or set of values, and finally the very specific criteria for the design that follows the norms. Van de Poel points out that it is in the norm segment that most of the arguments are taking place in order to ensure that the value/value set is right, and that the criteria for the design meets the norm requirements.

Based on extensive stakeholder analysis we find that trust and honesty are central to the majority of stakeholders (the example is fictional). The two values are placed in the value segment at the top of the model. To realize these values, the job is now to argue for the norms that are necessary to ensure that the values are visible, with an explicit explanation of the design/technology. Such standards could be: transparency, accessibility, recognition, and inclusion. These standards, from a VSD perspective, have been tested on stakeholders and on the set of values that underlie specific criteria for design. In telecommunication technology solutions this refers to the criteria, for example, for recognition (supporting confidence), i.e., that for instance the citizen does not feel alienated in relation to the telecommunication technology solution. But this is just one aspect of trust. It means that the technology must be reliable, which often implies that it must be mature, and that it should be open in the way it communicates information and knowledge to the recipient. The sender must be able to follow both the data and the way the data are treated.

In the development of the design criteria (technical specifications) other stakeholders in the VSD process will likewise take part. It is not readily open to laymen (aside from the fact that they are heard in the process), so it is our recommendation that VSD cannot stand alone. The focus is on interactional expertise in multidisciplinary teams, and it is therefore relatively closed toward the practical, everyday knowledge and experience that citizens, patients, families, and informal carers possess. Taking the value empowerment as an example for VSD approach, the following model is made up (Fig. 6.7).

The norms are chosen on the basis of both the system-world and life-world. Self-management and expertise ensure the system's interests, while life-world and equality are related to the experiential and moral standards. Criteria for design points toward what is needed to meet the norms. These are not specified in detail on technical measures, which would be the next stage in the model. The above might suggest that to achieve a technical specification level, the model should be denser and therefore have more steps.

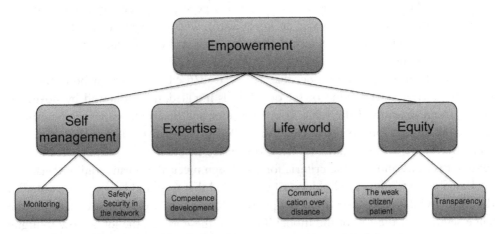

Figure 6.7 Model for value sensitive design with empowerment as point of departure.

VSD opens up an interdisciplinary approach whereby disciplines work closely together at all levels as indicated by VSD: conceptual studies, empirical studies, and technical studies. In this way VSD separates itself from the multidisciplinary, where several disciplines work together on a common project, but merely contribute with the expertise of the professionalism offered and do not interfere in how the knowledge presented is used and implemented. An example of the latter would be where the philosopher defines the value (concept), the sociologist/anthropologist makes the empirical studies, and engineer designs the artifact. In an interdisciplinary VSD project, academic skills would interact at all levels and exchange knowledge and expertise for the project's optimum implementation, in which values and norms are respected.

We have shown that from a telecommunication technological perspective, it is necessary to relate to the values: empowerment, emancipation, and enhancement, and that these values should be activated in relation to all direct and indirect stakeholders in the development and implementation of telecommunication technology. This means that norms and design criteria are complex in nature and, as Manders-Huits points out, there may be a need for a "value advocate" who manages to create hierarchy, structure, and operability. Manders-Huits recommends a specific and explicit ethical theory and a concrete person for completing the advocating role. We are, however, fairly critical of this model, as we believe it is too narrowly and top-down controlled, and also may block the exchange and development of ideas and practices in between stakeholders in the two-sided market.

Participatory studies of how social networking technologies and telecommunications technology may enhance compliance in health

User involvement in the development and use of social networking technologies

Social technologies are technologies that promote network relationships among their users. Social networking technologies can supplement and support treatment of chronic diseases. Social technologies may, if designed and adapted to the target group (e.g., people with a chronic illness and their caregivers), affect the target group's relationships and activities. If the right support is given, citizens' quality of life, perception, and outlook on their own health situation can be nudged, and citizens made to live with their health challenges in a positive and socially sustainable way.

However, for this to happen it is necessary to approach citizens with chronic health issues as competent and knowledgeable people despite their social, economic, and health literacy capabilities (Storm, 2012).

This section is intended to raise awareness of how social networking technologies have the potential to support both formal and informal networks of chronically ill citizens who are being treated for their diagnosis, and as part of this, using telehealth in the home.

Analyses of patient networks, e.g., at sundhed.dk, proved that patient network facilitates empowerment of the individual patient (Wentzer and Bygholm, 2013). Therefore, if the goal related to the use of tele-technologies in healthcare is to assist citizens in becoming active actors for their own health through participation in patient networks, and not only to share with each other the information they have received from health professionals, there is need to apply user-involving methods. These participatory methods can support the involvement of citizens in the design and implementation of the technologies of healthcare processes that citizens—as end-users—must be future players.

Studies from England show that citizens like to help other citizens with specific activities (Seyfang, 2003). In Denmark, the Red Cross has great success with visitors services, and in many parts of the country there is a widespread self-organized citizen network that helps refugees to start a new life as Danish citizens.

User involvement

The use of social technologies with the aim of promoting networking among citizens requires that those who are expected to participate in the networks will be involved when the technology is designed and implemented. Therefore, user involvement in healthcare is about creating spaces to enable citizens to contribute and innovate at an early stage in the technology design process. There is a general consensus that it is very good to involve patients and citizens in the health sector activities. This debate is taking place in the press and in public politics. A formal publication was published in April 2015 by the Danish Regions in the Municipality association and all the health professionals' organizations: "Joint Statement: Citizens' healthcare—our health."

> *"We focus on the life that people want to live outside the healthcare system and how this increasingly should be integrated into the treatment, care, and rehabilitation process. Therefore, we are talking about citizens rather than patients. The focus is on citizens who are in contact with the healthcare system and therefore have specific experience with health services, service, and culture."*
> **Danske Regioner, 2015**

Their vision is clear:

> *"We want to promote a culture in healthcare where we create safety, quality, and optimal processes, and involve citizens in decisions about their health and treatment. A culture in which citizens' knowledge, needs, and preferences are recognized and central to treatment, process, and organization. We must, because it gives better results to citizens, increase quality of life and improve the quality of health interventions."*
> **Danske Regioner, 2015**

There is consensus among health professionals' organizations and local, regional, and national health actors—across sectors and professional differences—that it is important that citizens (end-users) have a prominent position where healthcare is providing its core services.

It is important here to note that health professionals do not only play an important role in supporting citizens and their relatives in the use of various forms of tele-communications technology in the intended manner. Health professionals also have an important role when it comes to motivating, nurturing, and supporting citizens' use of social network technologies, as these technologies can serve as supports to sustaining a high quality of life when, and in spite of the fact that, life is being lived with a chronic diagnosis as a companion.

There are various approaches and methods for how to get insight in citizens' knowledge, needs, and preferences, so that citizens actually achieve a say in their own treatment. Fig. 6.8 below outlines three different approaches to user involvement in technology development. These have been selected because they can all assist in increasing the understanding of approaches to working with citizen participation in design, development, and implementation of both social technologies and tele-technologies in healthcare. In the first approach to user involvement, users inform designers and developers, and may also be used to test technologies (A). The second focuses on a mutual cooperation process between the users and the technology designers/developers (B), and the third is predominantly user-driven. Here users design through processes where multiple users create concepts and directions for designers of new technologies (C).

User-centered design process

Traditional technology design involves users only in the late stages of the design process, where they are used to evaluate or test design ideas and thereby makes suggestions for changes to the design. User-centered design is a broad term covering design processes where end-users indirectly affect how a design takes shape.

User-centered design seeks to embrace the needs and desires of the user and thus view the user as someone that has important inputs to contribute. This type of design became widespread after Donald Norman and Stephen Draper in 1986 published the

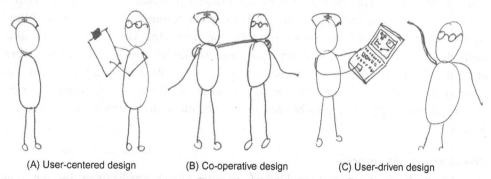

(A) User-centered design   (B) Co-operative design   (C) User-driven design

Figure 6.8 Different approaches to user involvement in technology development (A. Kushniruk and Nohr, 2016).

book, "User–Center System Design: New Perspectives on Human–computer interaction" (Norman and Draper, 1986). In this and a subsequent publication, "The Design of Every Day Things" designers are encouraged to study users and take account of their needs when designing new products. The challenge for the user-centered design approach is for the designer to understand user needs and to design technology to meet these needs.

The user-centered approach thus recognizes that users have something valuable to contribute to the technology design process and the approach can be characterized by the developer or designer observing their behavior and taking note of (1) what the users are doing or not doing, (2) what they prefer or do not prefer, (3) what they interact with and (4) what they have a need for. For this purpose various methods can be used, such as observation, (e.g., participant observation, video observation), personas, scenarios, use cases, usability tests, interviews (semi-structured or open), and questionnaires.

Cooperative design process

The cooperative design approach aims to involve various stakeholders in the design process to ensure that the outcome meets the needs of the users and the stakeholders of given context and thus that the designed product is useful.

The codesign approach arose from the recognition that the use context can be complex as well as difficult to access, and involve human activities including coordination and cooperation among many individuals with different areas of expertise. When the number of people involved in a use context exceeds a few, the complexity increases and the need for coordination increases. Computer Supported Cooperative Works (CSCW) is a key research area behind this approach. It works with how multiple-stakeholders cooperate and coordinate their activities and how this can be supported by computer systems (Carstensen and Schmidt, 1999). A key challenge in this approach is the way in which cooperation with the users in the design process is created and supported. This includes setting up a design processes where both users and designers are involved and where the aim is to create common designs and modifications to designs (Bødker and Grønbæk, 1992, pp. 199–218). Besides using the same methods as described above, prototypes are used to display and simulate future work situations in Living labs, or in situ simulations (as described in detail above in this chapter). The use of, e.g., prototyping and simulation, allows the participating users to explore and experience possible future use situations and discuss these with the designers.

User-driven design process

User-driven design (Participatory Design) requires direct involvement by the user throughout the technology design process. This approach originates from the social and political movements in the 1960s and 1970s. Participatory design of information technology started in Europe and Scandinavia but was soon appropriated by the United States as well. It initially focused on cooperation with the workers and trade unions on enhancing their participation in the development of new technology in the workplace (Robertson and Simonsen, 2013). The emphasis on user-driven designs is to anchor the design of technology in users' input and perspective, and to place the users in the front seat, actively involved in decision-making and the development of design concepts.

User-driven innovation is a methodology where the basic idea is to engage the user to innovate and develop design concepts for technologies. The users are the key players and the key aspects of the design decisions come from them. The designer's role is to plan and facilitate the processes to look forward to creating space for users to design. von Hippel has systematically studied this and recommended that user-driven innovation projects focus on "lead users" as the primary source of innovation (von Hippel, 2005). A specific step-by-step method for designing with users in innovation processes has been developed, and described in "User Innovation Management" (Kanstrup and Bertelsen, 2011). The UIM method aims to create a space for users to engage in innovation processes that can show the way for the designers. The method consists of three concepts (cooperation, context, and concepts) and six stages (select, plan, insight, vision, sketch, and present) and a number of techniques that facilitate the users to innovate.

UIM is based on the recognition that it is not possible—or desirable—to know the direction of the user's design process before you start. The purpose of going through a user participation process is that the designer along the way learns to find the direction of what the designated users need.

Challenge

The challenges that meet the tele-technologies in the healthcare user environment, when the results have to be evaluated and the health sector assesses the outcome, will affect the policy agenda. Technology designers should see technology users as active participants who have knowledge to contribute.

Citizens need support to remain in their own homes as long as possible and social network technologies have great potential to support people with chronic diseases, if developed and configured to and with the target groups.

The design of these types of tele-technologies needs to be based on how people want to live with these technologies—be it when people are alone, but also very much when they are with others, whether it is like-minded citizens/patients, or friends and family.

It is a challenge to develop and support new social networks for users of the future of tele-technology in healthcare.

Initiatives aim toward making it possible for citizens to perceive their chronic diseases as: "I am diagnosed with a chronic disease (e.g., COPD) and therefore I live my life in consideration of this," and not; "I am diagnosed with COPD therefore my whole life is COPD." Together, tele-technology in healthcare and social networks technology go hand in hand to create the opportunities for this and similar visions to be supported. Only the users of tele-technology in healthcare know how it is to live with a chronic disease.

CONCLUSION

In the methodology section of this chapter, we propose a triangulation involving simulation, values, and participation. This triangulation is based on the research that the authors of this chapter are representatives of and is therefore not an absolute set of methods to cover issues relating to the introduction of telecommunication technology in healthcare. Other and equally relevant methods could be used, but we believe that many of the key issues of empowerment, emancipation, and enhancement are illuminated through the use of the three methods.

Methods are characterized by acting as cones of light that illuminate specific parts of a reality, while other parts of the same reality remain in darkness. Clinical simulation illustrates the everyday practices in clinical reality, where one pragmatically seeks to achieve solutions that work best in the specific context. However, clinical simulation says very little about how the citizen/patient should be seen and treated in the given context.

Value-sensitive design focuses on fundamental human values (in a Western context) and highlights the dilemmas we face when we design for such a complex reality as healthcare. In an attempt to avoid an inappropriate philosophical abstraction level that blocks the application in practice, VSD works with technical design criteria. Our review of VSD shows how difficult an exercise this can be. VSD also suffers from the weakness that it does not relate to the technologies' potential agency, and exclusively focuses on actors and stakeholders, i.e., humans.

Participatory design or involvement of social groups in the design and implementation processes focuses on citizen/user/patient's needs, demands, and wishes in a dynamic process where there is a recognition that these are not absolute and static. Therefore, needs, demands, and desires are constantly evaluated in relation to the citizen/user/patient's physical and mental state. The method suffers from the fundamental weakness that it has a tendency to turn an "is" into a "must." By this we mean that if we design based on a specific condition in a presence and idealize this in standards and procedures, we risk blocking innovation and improvement.

If we look at the triangulation as such, we believe it is strong in relation to uncovering practices, values, needs, desires, and requirements for all directly or indirectly involved and affected actors and stakeholders in the design and implementation of telecommunications technology in healthcare. At the same time we recognize that the methods tend to explain historical and contemporary aspects of human-technology relations, with a focus on the human being, which may hamper innovation and change on the technical and technological level. Therefore, it is necessary that these dark areas are illuminated by complementary methods of a more technical nature, which in turn are revolved toward the creation of multidisciplinary and/or interdisciplinary teams that work together in a holistic, viable, and sustainable way, so that telecommunications technology solutions support better care and better health at less money.

This chapter has a partial two-sided market approach where we have focused on consumers constituting the two sides; where technological solutions should mediate appropriate and "catalyst" solutions for engagement and interaction. We have identified certain confusions in the terminology that need to be addressed in order to cope with the nature of this boundary object, where several professions, different types of knowledge and practice, and finally divergent interests, are at stake. We have introduced central values that we think are crucial in order to meet the requirements, needs, and wishes of consumers in a two-sided market, i.e., empowerment, emancipation, and enhancement.

ACKNOWLEDGMENTS

We wish to thank Gorm Simonsen, Region Nord (DK) and EU Social Funding for making this research possible. We also wish to thank Line Dausel Vinther, Master student at Techno-Anthropology (AAU), who has assisted us in every way during the research.

REFERENCES

Aalborg, K., 2014. Fremtidens Plejehjem. <http://www.fremtidensplejehjem.dk> (retrieved 27.10.15.).

Alpay, L., van der Boog, P., Dumaij, A., 2011. An empowerment-based approach to developing innovative e-health tools for self-management. Health Inform. J. http://dx.doi.org/10.1177/1460458211420089.

Ammenwerth, E., Hackl, W.O., Binzer, K., Christoffersen, T.E., Jensen, S., Lawton, K., et al., 2012. Simulation studies for the evaluation of health information technologies: experiences and results. Health Inf. Manag. J., 41(2), 14–21.

Boer, L., Donovan, J., Buur, J., 2013. Challenging industry conceptions with provotypes. CoDesign 9 (2), 73–89. http://dx.doi.org/10.1080/15710882.2013.788193.

Bødker, S., Grønbæk, K., 1992. Design in Action - from prototyping by demonstration to cooperative prototyping. In: GreenBaum, J., Kyng, M. (Eds.), Design at Work: Cooperative Design of Computer Systems. L. Erlbaum Associates Inc, Hillsdale, NJ, USA.

Brodersen, S., Lindegaard, H., 2015. The smart floor: how a public-private partnership co-developed a heterogeneous healthcare technology system. In: Bertelsen, P., Botin, L., Nøhr, C. (Eds.), Techno-Anthropology in Health Informatics. IOS Press, Amsterdam.

Carroll, J.M., 2000. Five reasons for scenario-based design. Interact. Comput. 13 (1), 43–60. http://dx.doi.org/10.1016/S0953-5438(00)00023-0.

Carstensen, P.H., Schmidt, K., 1999. Computer supported cooperative work: new challenges to systems design. In: Itoh, K. (Ed.), Handbook of Human Factors. Asakura Publishing, Tokyo, pp. 619–636. http://dx.doi.org/10.1.1.43.5157.

Commission, E., 2012. eHealth Action Plan 2012–2020 - Innovative healthcare for the 21st century (pp. 2–14). http://dx.doi.org/SWD(2013)527.

Cummings, M.L., 2006. Integrating ethics in design through the value-sensitive design approach. Sci. Eng. Ethics. Vol. 12, 701–715. http://dx.doi.org/10.1007/s11948-006-0065-0.

Dahl, Y., Alsos, O.A., Svanæs, D., 2010. Fidelity considerations for simulation-based usability assessments of mobile ICT for hospitals. Int. J. Hum. Comput. Interact. http://dx.doi.org/10.1080/10447311003719938.

Danske Regioner, M., 2015. Plan for Borgernes Sundhedsvæsen – vores sundhedsvæsen. Retrieved from <http://www.rm.dk/siteassets/om-os/aktuelt/borgernes-sundhedsvasen/plan-for-borgernes-sundhedsvasen_2015.pdf>.

Department of Health. (2011). Whole Systems Demonstrator Programme. Headline Findings - December 2011.

Evans, D.S., Schmalensee, R., 2013. The Antitrust Analysis of Multi-Sided Platform Businesses. Oxford Handbook on International Antitrust Economics, 623. (December). http://doi.org/10.3386/w18783.

Fumagalli, L.P., Radaelli, G., Lettieri, E., Bertele', P., Masella, C., 2014a. Patient Empowerment and its neighbours: clarifying the boundaries and their mutual relationships. Health Policy (Amsterdam, Netherlands) 119 (3), 384–394. http://dx.doi.org/10.1016/j.healthpol.2014.10.017.

Gadamer, H.-G., 1996. The Enigma of Health: The Art of Healing in a Scientific Age. Stanford University Press, Stanford, CA.

Habermas, J., 1968. Technology and science as "ideology." In: Seidman, S. (Ed.), On Society and Politics A Reader. Beacon Press, Boston, pp. 237–265.

Habermas, J., 1971. Knowledge and human interests: a general perspective. In: Shapiro, J.J. (Ed.), Knowledge and Human Interest. Beacon press, Boston, pp. 301–317.

Habermas, J., 2005. Teknik og videnskab som "ideologi." DET lille FORLAG.

Haraway, D., 1991. A Cyborg manifesto: science, technology and socialist-feminism in the late twentieth century. In: Harraway, D. (Ed.), Simians, Cyborgs and Women: The Reinvention of Nature. Free Association, London, pp. 149–181.

Jensen, S., Lyng, K.M., Nøhr, C., 2012. The role of simulation in clinical information systems development. Stud. Health Technol. Inform. 180, 373–377. http://dx.doi.org/10.3233/978-1-61499-101-4-373.

Jensen, S., Nohr, C., Rasmussen, S.L., 2013. Fidelity in clinical simulation: how low can you go? Stud. Health Technol. Inform. 194, 147–153. http://dx.doi.org/10.3233/978-1-61499-293-6-147.

Jensen, S., Rasmussen, S.L., Lyng, K.M., 2014. Evaluation of a clinical simulation-based assessment method for EHR-platforms. http://doi.org/10.3233/978-1-61499-432-9-925.

Jensen, S., Kushniruk, A.W., Nøhr, C., 2015. Clinical simulation: a method for development and evaluation of clinical information systems. J. Biomed. Inform. 54, 65–76. http://dx.doi.org/10.1016/j.jbi.2015.02.002.

Kanstrup, A.M., Bertelsen, P., 2011. User Innovation Management. Aalborg Universitetsforlag.

Kjeldskov, J., Skov, M.B., Stage, J., 2004. Instant data analysis: conducting usability evaluations in a day. In: Proceedings of the third Nordic conference on human–computer interaction, pp. 233–240. <http://doi.org/10.1145/1028014.1028050>.

Kragelund, L., 2013. The obser-view: a method of generating data and learning. Nurse. Res. 20 (5), 6–10. http://dx.doi.org/10.7748/nr2013.05.20.5.6.e296.

Kushniruk, A., Nohr, C., 2016. Participatory design, user involvement and health IT evaluation. In: Ammenwerth, E., Rigby, M. (Eds.), Evidence-Based Health Informatics. IOS Press, pp. 139–151. http://dx.doi.org/10.3233/978-1-61499-635-4-139.

Kuwata, S., Kushniruk, A., Borycki, E., Watanabe, H., 2006. Using simulation methods to analyze and predict changes in workflow and potential problems in the use of a bar-coding medication order entry system. AMIA ... Annual Symposium Proceedings/AMIA Symposium. AMIA Symposium, 994.

Lawton, K., Binzer, K., Skjoet, P., Jensen, S., 2011. Lessons learnt from conducting a high fidelity simulation test in health IT. Stud. Health Technol. Inform. 166, 217–226.

Lupton, D., 2013. The digital cyborg assemblage: Haraway's Cyborg theory and the new digital health technologies The Handbook of Social Theory for the Sociology of Health and Medicine. Palgrave Macmillan, Houndmills. 1–15. http://dx.doi.org/10.6084/m9.figshare.709639.

Manders-Huits, N., 2011. What values in design? The challenge of incorporating moral values into design. Sci. Eng. Ethics. 17 (2), 271–287. http://dx.doi.org/10.1007/s11948-010-9198-2.

Molin, L., 2004. Wizard-of-Oz prototyping for co-operative interaction design of graphical user interfaces. Nordic Conf. Human–Computer Interact. 82, 425. http://dx.doi.org/10.1145/1028014.1028086.

Newman, S.P., 2014. Terminology - Assistive Technologies. Retrieved from <http://2014.e-sundhedsobservatoriet.dk/sites/2014.e-sundhedsobservatoriet.dk/files/slides/StantonNewman,P1,web.pdf>.

Nielsen, K.G., Sabroe, C.D., Larsen, J., 2014. Patient Empowerment - Fra Strategi Til Handling. Aalborg University.

Nøhr, C., Villumsen, S., Bernth Ahrenkiel, S., Hulbæk, L., 2015. Monitoring telemedicine implementation in Denmark. Stud. Health Technol. Inform. 216, 497–500.

Pols, J., 2012. Care at a Distance - On the Closeness of Technology. Amsterdam University Press, Amsterdam.

Rasmussen, S.L., Lyng, K.M., Jensen, S., 2012. Achieving IT-supported standardized nursing documentation through participatory design. In Studies in Health Technology and Informatics, Vol. 180, pp. 1055–1059. <http://dx.doi.org/10.3233/978-1-61499-101-4-1055>.

Rasmussen, S.L., Jensen, S., Lyng, K.M., 2013. Clinical simulation as a boundary object in design of health IT-systems. Stud. Health Technol. Inform. 194, 173–178. Retrieved from <http://www.ncbi.nlm.nih.gov/pubmed/23941951>.

Region Hovedstaden, Center for Telemedicine (2013). Kortlægning af telemedicinske initiativer i Region Hovedstaden. Copenhagen.

Robertson, T., Simonsen, J., 2013. Participatory design, an introduction. In: Handbook of Participatory Design, p. 294. Retrieved from <http://books.google.com/books?id=SnO5JDzp3t4C&pgis=1>.

Seyfang, G., 2003. With a little help from my friends." Evaluating time banks as a tool for community self-help. Local Economy 18 (3), 257–264. http://dx.doi.org/10.1080/0269094032000111048c.

Storm, I.M.S., 2012. Når patientuddannelse øger ulighed i sundhed. Sygeplejersken 4, 90–93. Retrieved from <https://dsr.dk/sygeplejersken/arkiv/sy-nr-2012-4/naar-patientuddannelse-oeger-ulighed-i-sundhed>.

von Hippel, E., 2005. Democratizing Innovation. The MIT Press, Cambridge, MA, pp. 19–31.

Wentzer, H.S., Bygholm, A., 2013. Narratives of empowerment and compliance: studies of communication in online patient support groups. Int. J. Med. Inform. 82 (12), e386–e394. http://dx.doi.org/10.1016/j.ijmedinf.2013.01.008.

CHAPTER 7

Implementation and Evaluation of E-Health Ecosystems in Two-Sided Markets

P. Nykänen
University of Tampere, Tampere, Finland

INTRODUCTION

In the ICT-domain an ecosystem refers to a socio-technical system that is composed of domain organizations, actors, individuals, and technology-mediated communication, means and systems that in collaboration provide value, information, and services (Neely and Kastalli, 2013). Ecosystems are often complex, internal relationships are strong and the system develops around a solution, service, or technology. An ecosystem starts to develop when the actor who has developed the solution builds part of the value chain on the collaborators' support and manages the final value provision to the customer or to the end-user. The ecosystem develops and grows when new partners—actors—join in, who from their own part also support the value chain. New actors/partners are willing to join because they get benefits of being partners in the ecosystem, e.g., better quality, improved efficiency, or some other benefit valuable for them. The bigger the benefit, the more interesting the ecosystem is for the collaborators, and the stronger it will become. These qualities make the ecosystem interesting for the customer or consumer and this in turn adds to the strength of the ecosystem. When an ecosystem involves many partners, actors, they all have their own roles and tasks as collaborators, or as subcontractors, or even as competitors. In an optimal ecosystem the partners are searching for the best solution for the customer in such a way that all partners receive their expected benefits, communication inside the ecosystem is open and constructive, and collaboration is planned and follows the agreed ways of action. This kind of ecosystem is competitive and also creates new business possibilities in the long run (Messershmitt and Szypersky, 2003; Neely and Kastalli, 2013).

We use the term e-health in the meaning defined in Eysenbach (2001) as an umbrella term for an emerging field in the intersection of medical informatics, public health, and business, referring to health services and information delivered or enhanced through the Internet and related ICT technologies. In a broader sense, the term e-health characterizes not only technical development, but also a way of thinking

E-Health Two-Sided Markets.

and a commitment to a networked, global approach to improving healthcare locally, regionally, and worldwide using ICT. E-health has the potential to provide substantial benefits to both personal health and public health. It may empower the individuals in self-monitoring, in chronic disease management, and through providing access to trusted health knowledge sources. It also improves the ability to support surveillance and management of public health interventions and to analyze and report on population health outcomes (ITU and WHO, 2012).

E-health ecosystems are dynamic systems that evolve over time and incorporate varying numbers of stakeholders. Different types of collaborations are needed in real e-health ecosystem implementations. Due to the pervasive and evolving characteristics and requirements, the implementation and evaluation issues in the future will most likely be better dealt with by the ecosystem mode than the traditional health ICT system design and development mode. Ecosystems can produce the planned information systems in healthcare, where several collaborating actors form the synchronized value chain that delivers integrated service components that provide high quality healthcare outcomes and services. An e-health ecosystem involves different participants' roles which impact on the ecosystem stakeholders, such as citizens, health professionals, hospitals, health-related business actors, and governments. A specific feature of the e-health ecosystem is that it is strictly controlled by regulation and normative rules to protect the domain-specific features such as data security, privacy and access, and disclosure of health-related patient data.

To be able to evaluate the ecosystem, its performance and achievement of the planned benefits, evaluation is needed as an integral part of the ecosystem's lifecycle. Evaluation provides the means to assess the quality, value, effects, and impacts of the e-health ecosystem. Evaluation can be defined as measuring or exploring properties of e-health during the ecosystem lifecycle and the results of evaluation inform a decision to be made concerning the ecosystem in its context (Brender, 2006). Evaluation offers methods and tools to collect evidence about the achieved benefits, quality, effects, and impacts.

Evaluation studies have been performed since the 1960s, however most with a rather narrow scope, often focusing on how e-health systems are related to professionals' roles, the management of changes in their tasks, and user involvement (van Gennip and Talmon, 1995). Later studies have covered also the successful aspects and lessons learned in implementation and development of e-health systems. In many cases evaluation has been led by research interests to develop methodologies or to study the healthcare processes. From the 1980s onwards, management issues, user acceptance, and adoption of e-health systems in healthcare organizations have also been studied. Kaplan and Shaw made a review of how aspects related to people, organizational, and social issues have been considered in e-health evaluations (Kaplan and Shaw, 2004). They emphasize the need to pay more attention to these issues during e-health system design, implementation, and use, and emphasize the need to integrate multimethod evaluation to the whole lifecycle of the e-health system.

Ecosystem evaluation is an emerging field of interest and especially interesting is evaluation in two-sided markets, due to the complexity of the situation and the lacking methodological frameworks for evaluation. We discuss these issues in this chapter and summarize by presenting an integrative approach for ecosystem evaluation. The chapter is based on the recent published literature sources, and also on our ongoing research project, related to an ecosystem implementation and evaluation in the wide-area complex healthcare information system procurement process (Nykänen et al., 2016).

CHALLENGES FOR SUCCESSFUL E-HEALTH ECOSYSTEM IMPLEMENTATION

An e-health ecosystem is composed of healthcare organizations, both public and private, service provider professionals, customers, citizens, and patients, industrial companies providing their products and services, and technology-mediated communication and infrastructures that in collaboration provide add-on value for both service consumers and other service providers. Infrastructures and networks are needed both for knowledge sharing and management and for exchanging and communicating information and data. An e-health ecosystem in two-sided market means that services and products are developed and delivered to fulfill the customers' needs, or regulation-stated needs and the role of marketing and commercialization is minor (Vimarlund and Mettler, this book). Customers are both health professionals and patients, citizens, and service providers are healthcare organizations and various suppliers and industrial companies to provide systems and services to be used by healthcare organizations and patients. Examples of e-health applications and services of an ecosystem are electronic health records, personal health records, patient portals, health information systems including health knowledge management and e-learning for healthcare professionals, clinical decision support systems and remote patient monitoring and management as well as many wellness and fitness applications to be used at home and on the move by patients and citizens.

In the e-health domain an ecosystem consists of a number of actors of different types that are connected to each other in multiple and varying ways. Healthcare organizations and their requirements form the core of the health service system. The ecosystem members provide digital services, but do not necessarily provide a complete service for consumers, and instead can just render a specific part of a composite service. However, healthcare systems and service development are affected by many other actors that provide requirements for the functionality and performance of the system. Physicians and nurses are the users of such systems and their everyday practices need to be taken into account. Patients have traditionally experienced the systems indirectly, but the existing technological alternatives, the current paradigmatic change to citizen-centered care, and the need for patients' involvement and further improvement of the systems' efficiency provide better possibilities for the patients to be included in the ecosystem. Different kinds of vendor companies form business networks around the core system based on their core services and product offerings. Over time, the positions of the companies may change according to the exchange of resources in the business network (Adner and Kapoor, 2010; Serbanati et al., 2011, Lee, 2001).

We are missing today explicit methodological frameworks for ecosystem implementation, and therefore it is important to focus in ecosystem implementation on the following five key elements (Shehadi et al., 2011):

- *Governance policies and regulations*: The legislation policies and regulations need to address strong concerns over privacy and security, and to include security and confidentiality measures that assure all stakeholders that personal information will be protected. In creating a supportive legal environment for the e-health ecosystem we need to look at the four A's of sustainability: Authority (the power to effect change), Ambition (the desire for or intent to create improvement), Ability (the financial and human capital required for long term success), and Agility (the willingness to obtain feedback, observe opportunities, and adapt).

- *Financing model*: The appropriate funding is needed for the design, development, implementation, and ongoing operation of e-health ecosystems. Funding may come from many sources, e.g., the government or the public–private partnerships or insurance sources. The type of reimbursement model needs consideration as well as the incentives provided for stakeholders to encourage their participation in the ecosystem.

- *Technology infrastructure*: The selection of a specific technological platform has strong effects, it determines the applications, data, and infrastructure needed to support specific services and can realize fully the benefits of an e-health ecosystem. More important, applying open and publicly available, shared standards is crucial to making this platform ubiquitous.

- *Services*: E-health services should be tailored to local demands and to the available or planned technology infrastructure in order to ensure they meet customers' needs. Services may be as sophisticated as national electronic health records, or as simple as text-message alerts from public health entities to educate and inform patients on specific conditions, e.g., visit reservations to healthcare organization or measurement of the important blood glucose at home as part of diabetes monitoring. It might be wise to choose for the first implementation those services that are relatively easy to implement and endorsed by all parties; this would build a good start for the e-health ecosystem.

- *Stakeholders*: It is critical to have a human-centric approach and to involve carefully all key stakeholders from both the public and private sectors into an e-health ecosystem. Without the early support of all the key stakeholders and an alignment of their needs and objectives, the ecosystem may not be successful. Incentives for each group of stakeholders will have to be considered and identified.

Implementation of an ecosystem is a process with many phases (Messershmitt and Szypersky, 2003; Neely and Kastalli, 2013; Nykänen et al., 2016). From the beginning, it is important to include all the important players in the ecosystem. The most relevant players need to be identified on the basis of how crucial the ecosystem objectives are to them, and this can be done by interviewing with structured questionnaires. These interviews help to understand how the ecosystem needs to be organized and how it functions, as well as to understand the functions of each player. Interview analysis helps to identify the ecosystem roles that a player might adopt in helping to solve the ecosystem challenge. The priority of ecosystem objectives, and choice and prioritization of sub-objectives, will be heavily influenced by both legacy and emerging opportunities and needs.

The relevant phases of the ecosystem implementation according to (Neely and Kastalli, 2013):

- *Step 1—Identify the ecosystem objectives and opportunities*: Identify the complex challenges, objectives, and the solution. How could the objectives be broken down? What are the parts of the objectives and solution? How are different sub-objectives being solved? How is the comprehensive solution integrated from the sub-solutions?
- *Step 2—Analysis of the ecosystem*: Identify the players, their ecosystem roles and relationships. What are the rules of the game? What are the relationships, what do they contain? Identify the information flows. How does the data and/or money flow? Where does the power lie?
- *Step 3—Identify and implement the innovation opportunities*: Identify the gaps and how you can help to solve the problems. Find iteratively the best way to approach the solution. Introduce new or different services to solve the problem. Can you think of an even better way to solve the problem?
- *Step 4—Review and react*: Evaluate the progress, follow the development, and review and react to the continued evolution of the ecosystem.

Ecosystem requires enabling information, communication, and empowerment mechanisms which make it possible for information and expertise to be accessed quickly and accurately to inform and guide the ecosystem activities and performance. Possible mechanisms include: forums and dialogues, which facilitate broad engagement; expert communities, which facilitate expert engagement and capacity development; toolkits, which facilitate empowered learning and participation. Broad engagement of stakeholders in online and offline forums, social media, and dialogues is an important part of the emerging ecosystem. Expert communities offer good ways to share information related to expertise, e.g., tacit knowledge, experiences, interests, and organizational relationships (Serbanati et al., 2011).

Vimarlund and Keller (2014) made a literature review to find the prerequisites for successful implementation of IT-based innovations. They found that half (50%) of the reviewed implementation studies were performed without any guiding theory, model, or framework. This finding explains largely why many implementation projects fail. They summarized as successful factors for implementation: creation of common understanding of the implementation process among stakeholders; applying professional project and resource management; having strong guiding coalition; offering to end-users added value through the change caused by implementation (Vimarlund and Keller, 2014). They also found that there was a gap between the knowledge of how to carry out an IT-based innovation implementation and how these implementations have been received by organizations, users, consumers, and stakeholders. These findings are very relevant for the e-health ecosystems, too (Bibin, 2014; ITU-T, 2014). However, in the e-health domain, an ecosystem approach may help to overcome problems in implementation as ecosystems help to involve end-users and other stakeholders in the implementation process thus contributing to planned effects and impacts achievement, and helping to avoid nonexpected effects. Ecosystem approach also provides a framework for implementation, a guiding framework where the selected implementation methodology can be executed and followed. More research is, however, still needed with the ecosystem implementation frameworks and methodological approaches.

An e-health ecosystem needs to be implemented to sustain the expected e-health services and this implies the implementation of the required ICT infrastructure. The ecosystem also needs to be flexible enough to evolve in line with the development of new ICT or services (ITU-T, 2014). Sustainability is important not only because the start-up costs of any ecosystem system are significant, but also because the benefits usually emerge only over a period of several years.

Challenges of e-health ecosystems in two-sided markets

Two-sided markets are defined as markets in which one or several platforms enable interactions between the end-users, and try to get the two sides to work together by appropriately benefiting each side (Rochet and Tirole, 2005). These kinds of markets involve two groups of agents who interact via the platforms, where one group's benefit from joining a platform depends on the size of the other group that joins the platform (Amstrong, 2006). In the e-health domain the two-sided market situation is complex as the patients and citizens, health professionals, and health service providers on the one hand and on the other hand the many industrial companies and suppliers providing tools and systems that enable the interaction. Both groups receive benefits, and the more benefits, the more likely that more users and service providers are interacting. In the e-health two-sided ecosystem the role of marketing and customization is minor today, but may grow when the private healthcare providers take a more powerful position in service provision. Also, customization may be stronger in the future as we are progressing toward personalized healthcare services and tailored medicines and treatments. E-health ecosystem is very specific in the sense that political decision making has strong effects on how the healthcare services are organized and funded, and also, there is strong legal and normative regulation on how the services can be produced, delivered, accessed, and charged.

What is then "the platform" in the two-sided e-health market? The platform in the literature (e.g., Rochet and Tirole, 2005; Amstrong, 2006) refers to an intermediate that creates value by enabling interactions between two, or more, customers. Thus the platform can be any means, e.g., information system, mobile communication channel, Internet portal, knowledge sharing space, that enables the interaction, and offers the service users the possibility to select services from various service providers using their own selection criteria such as quality, easy access, compliance with the needs, and cheap price.

An e-health ecosystem needs to address customers' engagement in various business model alternatives. Static business model templates and frameworks are not adequate for ecosystems, they are less suitable to analyze the nature of participation in an innovation ecosystem (Bahari et al., 2015; Weiller and Neely, 2013; Adner and Kapoor, 2010). In many e-health ecosystem services the aim is to drive patient engagement in the ecosystem and then the business model is often pay-for-outcome, not pay-for-service-use, like with the traditional healthcare services. When patients are wanted to be engaged, they will do so, if they find it beneficial for themselves, if they get positive outcome, e.g., a service that fulfills their needs, or a service that is accessible in the right time and the right place. One application example of these kinds of services are the patient portals, Internet-based services offered for patients by healthcare service providers. These portals have been recently extended to also cover many different wellness features, e.g., to enable patients to monitor their own activities, import data from other wellness or fitness applications, or to track, e.g., eating or training habits. More recently gamification has been included to these applications thus promoting user interaction and physical activity. The business model encourages the user adoption of these new services.

Today the e-health ecosystem needs to provide smart and personalized care and services for the patients and citizens. This can be achieved, e.g., with tools and systems that track patients' health at home. These kinds of systems need such user interfaces in their platforms that display the patient's health holistically or offer patient empowering features, and they can connect patient and physician communities and facilitate physician-patient interactions with new means and contents. The empowerment experience may exactly be the outcome that the patients are looking for and for what they are willing to pay.

Typical for an e-health ecosystem today is that e-health services, e.g., electronic health records, need to be delivered online, across distinct organizational or even across national borders. This requires that an ecosystem is composed of interconnected stakeholders, each one with a mission to improve the quality of care. In this situation the stakeholders build new relationships, often outside the healthcare organization. These relationships may become essential and cover the following (Bibin, 2014):

- *Provider–payer relationship*, both in public and private healthcare: The payer–provider collaboration is at the heart of the consumer- or citizen-centered model of healthcare. This collaboration may focus much today on wellness issues, instead of health. New collaborative care teams, or integrated care teams, may include representatives from payers and providers, these teams can learn how to tailor service to the consumer's needs.
- *Provider–pharmacist collaboration*: The healthcare providers may find the pharmacists as direct partners in patient care, e.g., monitoring the patient's adherence to prescriptions and delivering new information about adverse drug interactions. The pharmacist may have access to real-time data and a closer relationship with a patient's physician, this might save money and resources and potentially improve patient satisfaction.
- *Medical device manufacturers and health IT suppliers–clinician communication*: Clinicians will have to work closely with manufacturers and suppliers to develop new devices, applications, and e-health systems for themselves and their patients. The usefulness and usability of such systems will depend on user-centric approach in design and on application of open standards, interoperability, and communication.
- *Employer-payer relationship*: The cost structure and costs of healthcare will potentially change, cost-savings could be expected, when the patients have more freedom in selecting the service providers and many services can be accessed from home or on the move.
- *The consumer relationship*: As healthcare consumers, patients, and citizens become more aligned and integrated with care management they want to have intelligent and trustworthy tools and systems in use. E-health system developers should make tools available for self-monitoring, and provide consumers with the information and resources to make healthy lifestyle changes and choices. Trust, privacy, and security are key issues in building trusted relationships.

The ecosystem of public health is complex, it is characterized by a multiplicity of interactions among the numerous actors including, e.g., healthcare institutions, social services, educational institutions, social security services, governmental agencies. The large diffusion of the ecosystems in private healthcare may provide benefits for the public health ecosystem, e.g., for a global health surveillance service and through the use of anonymous personal health data (ITU-T, 2014). Today the e-health ecosystem in the healthcare environment is very much based on the public–private partnership, at least in the EU region, and on integrated care paradigm, where public and private health organizations and third sector service providers develop, together with industrial partners and suppliers, services that complement each other and have the purpose of fulfilling the regulative and users, patients and citizens, needs. This kind of public–private partnership ecosystem may have a very complex business model, however, the model should enable benefits to all ecosystem stakeholders (Lee, 2001). The benefits are not always measured with monetary measures, they may be e.g., empowerment experience, easy service access, user-friendliness of the service, or satisfactory outcome from the service use.

In the e-health ecosystem the consumer, the patient, the citizen has new power and prominence. All ecosystem stakeholders may need to rethink technology sharing and use of ICT to understand and improve consumer experiences (Bibin, 2014). And, while creating e-health applications in the context of the e-health ecosystem, customization and tailoring to users' needs are needed, each e-health system and effort needs to reflect its local conditions and culture.

EVALUATION OF E-HEALTH ECOSYSTEM IMPLEMENTATION

Evaluation is considered to be difficult as it deals with values and norms and various organizational contexts and stakeholders' interests, and it has to fight for funding and support. Additionally, evaluation results are to be analyzed and interpreted in the study context (Ammenwerth et al., 2004; Rigby et al., 2013). A challenge to improve the quality of evaluation studies is to apply a systematic approach, plan the evaluation study carefully and execute the study following systematic evaluation guidelines, e.g., the Good Evaluation Practice Guideline, GEP-HI (Nykänen et al., 2011).

As there are many focuses of interest for evaluation, e.g., economics, efficiency, usability, usefulness, safety, privacy and security, compliance with the clinical process and workflows, functionality, effects and impacts on healthcare outcomes, there is also need to use many different methods suitable for measuring the evaluation criteria of interest. These potential methods cover e.g., qualitative and quantitative methods, statistics, heuristics, ethnography, contextual inquiry, human–system interaction observations, data mining and quality analysis, cost-effectiveness, and cost-benefit analyses. There are also many potential perspectives for evaluation, representing various stakeholders' viewpoints, e.g., managerial, clinical, technical, industrial, governmental, and these viewpoints may be studied at various levels of healthcare system—local, regional, and national, or even at EU or international levels. It is of utmost importance, when planning an evaluation study, to elaborate and define how these issues are related to the study: Coverage of scientific robustness, relevance to the current purpose of the study, best fit of the important characteristics to the specific current need and relevance of the methods (Nykänen and Kaipio, 2016).

The quality of an evaluation study is dependent on many factors, e.g., on the objectivity of the study and on the independence of evaluators, referring to their independence on economic interests, on intellectual interests, and on the various stakeholders' interests. An evaluation study must also be scientifically well-established on robust theories and methodologies, and the study should be performed following the principles of scientific research. Our review (Nykänen and Kaipio, 2016) revealed quality problems in evaluation studies, e.g., weak study planning without a systematic, scientific methodology, false implicit assumptions made in the study, experimental errors in the research setting, under- or overinterpretation of the results or false conclusions, inclusion of non-neutral evaluators, intra- or interorganizational variability in the evaluation object or problems with the novelty of technology. In some cases there are problems in selecting such evaluation methods that are capable of measuring those variables and aspects that describe the phenomenon under study. Also in usability evaluations we found quality problems, e.g., the studies often focus on a single end-user group perspective, on user interface components, or on use of the system in a specific context, but do not provide a comprehensive picture of usability of a large-scale healthcare information system in use. Further, the evaluation studies rarely discuss the relationship between single-system development and the existing technology setting in healthcare, or the characteristics of various use contexts in which the evaluated system is used. Usability evaluation studies have in most cases been done separately from other evaluation studies, e.g., from effectiveness, effects, and impacts evaluation studies (Nykänen and Kaipio, 2016).

This state of affairs of evaluation studies and quality challenges emphasize the need for systematic approaches and guidelines to design and carry out different kinds of evaluation studies to provide evidence about the impacts and actual efficiency, quality, usability, and safety of e-health (Nykänen and Kaipio, 2016).

A crucial point with evaluation of an e-health ecosystem and its effects is that evaluation is to be performed from many perspectives; a single stakeholder's perspective is not any more sufficient. Most earlier evaluation studies have been focused on one user's individual context: How a health professional uses the system in the specific context, how he/she accepts the technology and how technology fits into his/her work processes, leaving the wider usage contexts untouched (Nykänen and Kaipio, 2016). With the ecosystems, this approach is no more applicable, ecosystems are dynamic, multiple evaluation perspectives and many issues of interest need to be covered in ecosystem evaluation.

The potential effects of e-health ecosystems should be assessed along the five key elements of ecosystems. From the governance policies and regulations perspective there is need to assess the effects to improve the regulations of privacy as the e-health ecosystems may enable ubiquitous health information spaces where dynamic privacy and security regulation are needed. We need to be sure that the EU level and national data protection and security rules are fulfilled, therefore evaluation needs to ensure that privacy and security controls are implemented to assure all stakeholders that personal information will be protected from unauthorized access and disclosure. From the funding model perspective the effects may be seen in strengthening of the public–private partnerships and thus effects on the partners. We need to evaluate how the design, development, implementation, and ongoing operation of e-health ecosystem is financed, and also how the healthcare services costs are reimbursed. The funding parties need to stay neutral in evaluation. From the technology infrastructure perspective the effects may occur in sustainable technology selections and in promoting use of standards. From the services perspective with the e-health ecosystem the effects need to be studied, e.g., from the following perspectives: How well the services adapt to the local customers' needs and improve the accessibility of the services. When developing ecosystem services it is important to apply a human-centric approach, and this means that evaluation has to be performed from the perspectives of all stakeholders and users to collect evidence on usefulness, usability, and accessibility of the services. From the stakeholders perspective the effects potentially are strong, all key stakeholders need to be involved both from the public and private sectors into an e-health ecosystem and they will need incentives that encourage them to use the services, and in this case the positive effects can be realized. Evaluation needs to be carried out from all relevant stakeholders' perspectives.

Many frameworks and guidelines for evaluation exist, aiming at supporting and improving evaluation studies so that e-health evaluation would be conducted to the highest methodological and scientific standards. These frameworks and guidelines differ in terms of generality, specificity, and timing, related to system development phases and theoretical underpinning.

Nykänen et al. (2011) have developed a Guideline for Good Evaluation Practice in Health Informatics (GEP-HI) which gives advice on how to design and carry out evaluation studies in various e-health contexts. The GEP-HI guideline lists issues to be considered at each evaluation phase, and gives recommendations on how to design evaluation studies, how to make methodological choices, how to conduct studies, and how to define evaluation criteria at a specific phase of the e-health lifecycle. The phases of GEP-HI guideline are (Nykänen et al., 2011):

- *Preliminary outline:* The purpose of the study and the first ideas on why, for whom, and how the evaluation should take place
- *Study design* clarifying the design issues for the evaluation study
- *Operationalization of methods* making the methodological approach and methods concrete and compliant with the system type, the organization, and the information needed
- *Project planning* developing plans and procedures for the evaluation project
- *Execution of the evaluation study* accomplishing the designed evaluation study
- *Completion of the evaluation study* reporting, accounting, archiving of evaluation study results, finalization of outstanding issues, and formal closure of the evaluation study (Fig. 7.1).

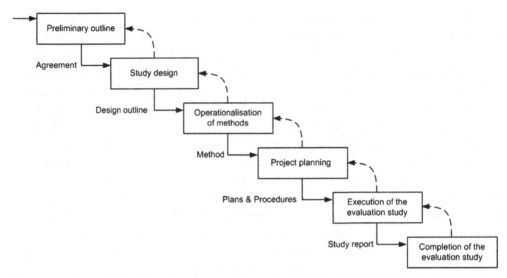

Figure 7.1 Phases of e-health evaluation according to the GEP-HI guideline (Nykänen et al., 2011).

The GEP-HI guideline can be applied in e-health ecosystem evaluation. It gives the systematic framework for the planning and execution of the evaluation study. However, the methodological choices need to be made depending on the evaluation study details, e.g., which evaluation criteria are of interest, which stakeholders are relevant for the case, what are the objectives of the ecosystem and what are the planned benefits that should be achieved.

Carroll et al. (2016) have recently developed a Connected Health Evaluation Framework (CHEF) for e-health ecosystem evaluation. They define Connected Health as a socio-technical healthcare model, which extends healthcare services beyond healthcare institutions. Thus Connected Health is seen to be an e-health ecosystem, as it builds a balance between the various requirements and dynamics associated with different stakeholder groups in a modern healthcare sector (Carroll et al., 2016). The CHEF framework is comprised of four main layers addressing clinical, business, users, and systems, and the purpose is to determine how these cocreate value. Each of the categories supports specific Connected Health operations across all service lifecycle stages, ultimately generating healthcare net benefits. CHEF provides a holistic view of a healthcare system, tailored analysis of healthcare service lifecycle, performance metrics on service operations and patient-focused analytics, scorecard and benchmark tools to assess healthcare technological integrations, healthcare interventions, and healthcare providers (Carroll et al., 2016).

Though many frameworks for evaluation exist, none of them is specifically planned to support ecosystem evaluation. An interesting challenge would be to integrate the CHEF framework with the GEP-HI framework in such a way that the

CHEF evaluation aspects: business growth, healthcare practice, end-user perception, are connected to the GEP-HI phases preliminary outline, study design, and operationalization of methods. The CHEF quality management aspect equals much with the GEP-HI execution of the evaluation study phase by referring to technical and regulative requirements and conformity assessment. The CHEF business growth focuses on cultural and strategy change, the healthcare practice focuses on e-health and innovation and how these change the practice and clinical pathways. End-user perception focuses on safety and quality of healthcare innovations from the user perspective (Carroll et al., 2016). These CHEF aspects, when combined with the GEP-HI phases, would help to define the focus of the planned evaluation study, to select the explicit criteria for ecosystem evaluation, and to carry out a systematic evaluation study.

Ecosystem evaluation must be conducted from many stakeholders' perspectives to assess the system's impacts across the broad spectrum of care services and their users. The scope of CHEF explicitly acknowledges the broad scope and existence of different stakeholders (Carroll et al., 2016). It seems therefore a promising approach to integrate CHEF with the GEP-HI guideline for e-health ecosystem evaluation. When integrating these two, the GEP-HI guideline provides a systematic approach with phases on planning and execution of an evaluation study, and CHEF framework introduces the aspects which address clinical, business, users, and systems with the purpose of determining how these cocreate value. These layers can be connected to the relevant GEP-HI phase. The layers represent the aspects of an ecosystem and the evaluation criteria and the methods to measure the criteria can be derived from these layers, depending on the scope of the planned study. This approach for ecosystem evaluation would support management of the evaluation and takes into account the various stakeholders' perspectives and different interests in evaluation.

DISCUSSION

Implementation of an e-health ecosystem is a dynamic process where all stakeholders and their perspectives need to be included. Ecosystems may be complex with strong internal relationships, and the system develops around a solution, service, or technology. In a sustainable ecosystem the partners are searching for the best solution for the customer in such a way that all partners receive their expected benefits, communication in the ecosystem is open and constructive, and collaboration is active. An e-health ecosystem involves different participants' roles which impact on the ecosystem stakeholders, such as citizens, research professionals, hospitals, health-related business actors, and governments. Specific to e-health ecosystem is that it is strictly controlled by regulation to protect the domain-specific features like data security, privacy and access, and disclosure of health-related confidential patient data.

There are not yet explicit models or methodological frameworks for e-health ecosystem implementation. A four-step approach for the development can be followed: (1) specification of the goals and objectives of the system, (2) identification of the stakeholders, analysis of the roles and tasks and competencies of the stakeholders, (3) identification and analysis of the solution possibilities, analysis of the strengths and weaknesses, and iteration to the final solution, and (4) evaluation and reaction to identified problems and challenges.

In e-health ecosystems, involvement of users in the whole ecosystem implementation and evaluation process is very important. Healthcare professionals, e.g., medical doctors and nurses, as well as citizens and patients as users of the services, need to be stakeholders in the e-health ecosystem. By involvement and participation they are able to improve their competencies, support in defining the scope of the ecosystem, in collecting the needs and requirements for the ecosystem, and in prioritizing these requirements. Additionally, health professionals as end-users are needed to describe the user scenarios and use cases and to participate in the evaluation activities. Through participation, health professionals will be committed to the ecosystem and its objectives, and they have less resistance to the solution provided by the ecosystem. Participation also builds up communities of practice, networks the professionals, users, and other stakeholders, supports their mutual communication, and prohibits them to be isolated from shared activities; all these support the sustainability of the ecosystem.

E-health ecosystems may apply many different business models. An e-health ecosystem is implemented to provide benefits for its stakeholders, like patients and health professionals. An example of this kind of service is the patient portals offering patients easy accessibility for the services and a new way to interact with health professionals. If patients adopt these kinds of services, the services most likely will be beneficial also to the health professionals, e.g., by saving them time and resources, by having better information on the patient's actual status and situation, and through the possibility to react faster on changes. Typical for e-health business models is that customers, patients, citizens, and healthcare organizations pay for the outcomes, not for the actions per se.

Despite the challenges in evaluation, the e-health ecosystem needs to be evaluated, during its implementation and maintenance, for the whole lifecycle. Ecosystem evaluation must be conducted from many stakeholders' perspectives to assess the system's impacts across the broad spectrum of care services and their users. The ecosystem evaluation needs an approach that supports management of the evaluation and takes into account the various stakeholders' perspectives and different aspects of interest in evaluation. By now, there are not many potential methodological approaches available for this kind of evaluation. We have proposed integration of the two frameworks presented in the literature, namely GEP-HI and CHEF, to cover the complexity of e-health ecosystem evaluation. So far, however, we have no experience how this integration will work in practice. More research is needed with e-health ecosystem implementation and evaluation.

CONCLUSIONS

E-health ecosystems are dynamic systems that incorporate a varying number of stakeholders. An e-health ecosystem in two-sided market means that services and products are developed and delivered to fulfill the customers' needs. Implementation of an ecosystem is a dynamic process wherein it is important to include all the important players and to identify their roles and relationships. Ecosystem requires enabling information, communication, and empowerment mechanisms which make it possible for information and expertise to be accessed quickly and accurately to inform and guide the ecosystem activities and performance. An e-health ecosystem is very specific in the sense that political decision making has strong effects on how the healthcare services are organized and funded, and also, there is strong legal and normative regulation on how the services can be delivered, accessed, charged, and funded. Typical for an e-health ecosystem today is that e-health services, e.g., electronic health records, need to be delivered online, across distinct organizational or even national borders. This requires that an ecosystem is composed of interconnected stakeholders, each one with a mission to improve the quality of care. To ensure in this situation the patient safety and quality care, the stakeholders build new relationships, often outside the healthcare organization.

In the healthcare clinical context, the e-health ecosystem consists of many health IT applications, of which several are used simultaneously. Evaluation of this kind of e-health ecosystem should address the relevant evaluation criteria from a broad viewpoint, from many stakeholders' perspectives. E-health systems and applications are part of the ecosystem and the objectives of evaluation should be framed with respect to the ecosystem objectives, stakeholders, information need and context. A challenge is to cover the wide variety of users and stakeholders of the e-health systems and the numerous purposes these systems serve, and the diversity of clinical surroundings in healthcare organizations where these systems are implemented and used. An integrative approach covering the frameworks GEP-HI and CHEF has been proposed in this chapter to address the complex evaluation situation.

The ecosystem approach on e-health systems and applications helps healthcare organizations and e-health stakeholders in creating business models by taking an ecosystem perspective on the outcomes they wish to create for users. The ecosystem approach has the potential to improve collaboration and communication between the stakeholders. The actors, partners of the ecosystem share the common objectives of the system and benefit from collaboration, create shared value, and when they have adequate tools and means for communication, communication and collaboration will be active and support the ecosystem sustainability. However, more research is needed to realize all the promises of the ecosystem approach.

E-health systems have many potential benefits for the healthcare professionals and organizations and patients and citizens and for e-health industrial suppliers. However, creating sustainable e-health ecosystems requires that all stakeholders' opinions and needs are taken into account for ecosystem success and sustainability. Sustainability is important because the benefits often can be achieved only over a period of time.

REFERENCES

Adner, R., Kapoor, R., 2010. Value creation in innovation ecosystems: How the structure of technological interdependence affects firm performance in new technology generations. Strateg. Manage. J. 31 (3), 306–333.

Ammenwerth, E., Brender, J., Nykänen, P., Prokosch, H.U., Rigby, M., Talmon, J., 2004. Visions and strategies to improve evaluation of health information systems. Reflections and lessons based on the HIS-EVAL workshop in Innsbruck. Int. J. Med. Inform. 73, 479–491.

Amstrong, M., 2006. Competition in two-sided markets. RAND J. Econ. 37 (3), 668–691.

Bahari, N., Maniak, R., Fernandez, V., 2015. Ecosystem Business Model design. XXIVe Conférence Internationale de Management Stratégique. <http://www.strategie-aims.com/events/conferences/25-xxiveme-conference-de-l-aims/communications/3522-ecosystem-business-model-design/download>.

Bibin, T., 2014. The new healthcare ecosystems – 5 emerging relationships. UST Global health group. <http://www.beckershospitalreview.com/hospital-management-administration/the-new-healthcare-ecosystem-5-emerging-relationships.html>.

Brender, J., 2006. Handbook of Evaluation Methods for Health Informatics. Elsevier Academic Press, USA.

Carroll, N., Travers, M., Richardson, I., Evaluating Multiple Perspectives of a Connected Health Ecosystem. In: 9th International Conference on Health Informatics (HEALTHINF), Rome, Italy, 2016, 21–33.

Eysenbach, G., 2001. What is e-health? Journal of Medical Internet Research 3 (2), e20. http://www.jmir.org/2001/2/e20.

ITU and WHO, National eHealth Strategy Toolkit, 2012, http://www.itu.int/pub/D-STR-E_HEALTH.05-2012.

ITU-T Focus Group on M2M Service Layer, M2M enabled ecosystems: e-health, ITU technical report, 2014, https://www.itu.int/dms_pub/itu-t/opb/fg/T-FG-M2M-2014-D0.2-PDF-E.pdf.

Kaplan, N., Shaw, N., 2004. Future directions in evaluations research: people, organizational and social issues. Methods Inf Med 43, 215–231.

Lee, C.-S., 2001. An analytical framework for evaluating e-commerce business models and strategies. Internet Research: Electronic Networking Applications and Policy 11 (4), 349–359.

Messershmitt, D.G., Szypersky, C., 2003. Software Ecosystem Understanding an Indispensable Technology and Industry. MIT Press.

Neely, A., Kastalli, V., 2013. Collaborate to innovate – How business ecosystem unleash business value, Cambridge service alliance. University of Cambridge., http://www.ifm.eng.cam.ac.uk/uploads/Resources/Collaborate_to_Innovate_-_ecosystems_-_final.pdf.

Nykänen, P., Kaipio, J., 2016. Quality of health IT evaluations. In: Ammenwerth, E., Rigby, M. (Eds.), Evidence-Based Health Informatics. Studies in health technology and informatics 222. IOS Press, Amsterdam, pp. 291–303.

Nykänen, P., Brender, J., Talmon, J., de Keizer, N., Rigby, M., Beuscart-Zephir, M.-C., et al., 2011. Guideline for good evaluation practice in health informatics. Int. J. Med. Inform. 80, 815–827.

Nykänen P., Tyllinen M., Lääveri T., Seppälä A., Kaipio J., Nieminen M., 2016, Guideline for an ecosystem implementation in an e-health procurement process. University of Tampere Research report 45 June 2016 (in Finnish), http://www.uta.fi/sis/reports/index/R45_2016.pdf.

Rochet J.C., Tirole J., Two-sided markets: a progress report, 2005, https://core.ac.uk/download/files/153/6634993.pdf.

Serbanati, L.D., Ricci, F.L., Mercurio, G., Vasilateanu, A., 2011. Steps towards a digital health ecosystem. Journal of Biomedical Informatics 44, 621–636.

Shehadi R., Tohme W., Bitar J., Kutty S., 2011. Anatomy of an E-Health Ecosystem. Booz & Company Inc. USA. <http://www.strategyand.pwc.com/media/uploads/Strategyand-Anatomy-of-E-Heatlh-Ecosystem.pdf>.

Rigby, M., Ammenwerth, E., Beuscart-Zephir, M.-C., Brender, J., Hyppönen, H., Melia, S., et al., 2013. Evidence based health informatics: 10 years of efforts to promote the principle IMIA Yearbook of Medical Informatics 2013. Schattauer Verlagsgeschellschaft, Stuttgart.1–13.

van Gennip, E.M.S.J., Talmon, J.T., 1995. Assessment and Evaluation of Information Technologies. IOS Press, Amsterdam.

Weiller C., Neely A., 2013. Business model design in an ecosystem context, University of Cambridge, Cambridge service alliance. <http://cambridgeservicealliance.eng.cam.ac.uk/resources/Downloads/Monthly%20Papers/2013JunepaperBusinessModelDesigninEcosystemContext.pdf>.

Vimarlund, V., Keller, C., 2014. The many faces of implementation VINNOVA Report VR 2014:06. Swedish Governmental Agency for Innovation Systems, Sweden.

Vimarlund V., Mettler T., The two-sided market of e-health and its ecosystem (this book).

CHAPTER 8

HIT Implementation and Coordinated Care Delivery from the Perspective of Multisided Markets

C.E. Kuziemsky
University of Ottawa, Ottawa, ON, Canada

INTRODUCTION

Healthcare systems worldwide are undergoing transformation to coordinated systems that enable the provision of collaborative patient centered healthcare delivery and the tracking of patient outcomes across providers and settings (IOM (Institute of Medicine), 2012). In fact, Bates calls the defining and establishment of care coordination the next great challenge for medical informatics (Bates, 2015). However, the road to achieving a coordinated and collaborative healthcare system is long and has many obstacles to achieving it. A foremost challenge is that coordinated care delivery shifts the landscape in which care is provided. As siloes are broken down as part of providing coordinated care delivery it presents new challenges such as healthcare providers supporting care delivery in areas where they have not traditionally provided care (Bates, 2015), e.g., surgeons or specialist providers who may now have to monitor patient care in the community. Another challenge is the need for individual providers to change workflows as part of providing collaborative care delivery (Kuziemsky, 2015).

Health information technology (HIT) will play a key role in transforming the healthcare system as it is the platform upon which coordination is based. One of the primary benefits of HIT is that it can establish connectivity between different users and processes. Early HIT such as decision support systems and order entry systems connected providers to evidence-based decision making and orders, respectively. As HIT evolved in functionality, the magnitude of coordination increased, such as supporting collaborative care teams at the meso level, and coordinating government level monitoring of costs and outcomes at the macro level. However, as the degree of coordination increases, so does the scope of issues from HIT-mediated coordination. The presence of unintended consequences post HIT implementation is well known, and the extent of these issues increases as the degree of coordination increases (Kuziemsky, 2016).

HIT can be a great support for care delivery models such as collaborative care delivery by providing resources such as awareness about the members of a patient's care team and the ability to develop and maintain a shared care plan across the team (Bates, 2015). However, it is presumptuous to assume that HIT on its own will improve care coordination. In fact, studies have highlighted how large-scale healthcare transformation initiatives predicated on supporting care coordination such as the Connecting for Health Initiative in the United Kingdom (Hamblin and Ganesh, 2007; McGlynn et al., 2008) and the Health Information Technology for Economic and Clinical Health (HITECH) Act in the United States (Blumenthal, 2011) have struggled to achieve their desired outcomes (Mennemeyer et al., 2016). HIT frequently leads to unintended consequences (UICs) post HIT implementation (Ash et al., 2007; Harrison et al., 2007; Kuperman and McGowan, 2011). These UICs occur because HIT creates connectivity across people, policies, and processes in ways that was not anticipated. While HIT such as electronic health records are designed to facilitate connectivity, to date the results of HIT-mediated connectivity have been mixed (O'Malley et al, 2010; Samal et al., 2013). Therefore, while care coordination may indeed be the next significant design challenge for HIT (Bates, 2015), we first need research to understand the coordination needs within and across different user groups and levels of care delivery.

One of the key challenges in establishing coordination is that different agents may have different priorities, leading to conflicts. For example, governments worldwide (i.e., macro level) are making a substantial investment in healthcare transformation and as such want better tracking of resources spent and care outcomes (e.g., wait times). However, the onus to conduct transformative healthcare processes and to collect front line data to measure care quality often falls on the front line providers (i.e., micro level), which results in unintended consequences because of a gap between the needs and priorities of macro and micro agents (Kuziemsky and Peyton, 2016; Vezyridis and Timmons, 2014). Individual providers (i.e., micro level) may find the transition to team-based collaboration (i.e., meso level) to be challenging due to the need to change individual practices to support collaborative needs (Kuziemsky, 2015). While studies have described a positive relationship between patient empowerment and satisfaction when patients use HIT (Rodriguez et al., 2011; Businger et al., 2007) we also know that it can lead to workflow concerns and other issues for providers (Lusignan et al., 2014).

The common factor in all the above examples is that HIT is the network that connects different agents within and across care delivery levels (i.e., micro, meso, and macro), but the different needs or priorities within and across the levels manifest themselves as HIT implementation issues. Coordinating care delivery across multiple levels is challenging for several reasons including the need to establish and maintain multiagency partnerships as part of delivering new approaches of healthcare delivery, and embracing innovative transformation while still respecting the scale and pace of day-to-day care delivery (Devlin et al., 2016). However, the reality is that HIT-mediated coordination is not solely a technological nor social endeavor but rather a sociotechnical one that must be studied from the perspective of both the technology and the social system where it will be used. We suggest that a better understanding of different implementation perspectives would enable us to better manage implementation issues.

This chapter presents a multisided market framework for coordinated healthcare delivery platform to provide insight on HIT implementation across multiple levels (e.g., micro, meso, and macro) of healthcare delivery. The framework helps us understand how different priorities and needs across the different levels present as HIT implementation issues. We then describe how the framework also allows us to better design and evaluate HIT to support different levels of coordinated healthcare delivery.

MULTISIDED MARKETS AND HEALTH INFORMATION TECHNOLOGY IMPLEMENTATION

A two-sided market approach is one in which two parties interact via a platform (e.g., information technology) and are dependent on each other for the success of the interaction (Rochet and Tirole, 2003). Credit card companies and social media sites are common examples of two-sided markets. Multisided markets are simply extensions where the market has more than two key players. Healthcare has been described as a multisided market given the multiple players involved in healthcare delivery (Mettler and Eurich, 2012). A multisided market perspective could be used to study HIT implementation as the three healthcare delivery levels (i.e., micro, meso, and macro) described above will only function successfully if the healthcare system as a whole works as an integrated and coordinated system. To achieve system-wide coordination we need to realize that the micro, meso, and macro levels rely on each other. At the micro level, patients want more involvement in their care delivery and providers need to effectively manage and coordinate patient care in order to provide effective and cost efficient care delivery. At the meso level, team-based coordination and collaboration relies on individual providers (i.e., micro level) doing their individual tasks while also participating in collaborative tasks across team members such as updating care plans (Collins et al., 2011). At the macro level, governments and healthcare delivery systems want care delivery to be accountable and economically sustainable and thus require data on care delivery processes and outcomes. Therefore, each level depends on coordination with the other levels for healthcare delivery to work effectively.

However, achieving this multilevel coordination is challenging due to different priorities across the levels. A multisided market perspective can be used to understand how HIT enables and constrains agents and processes across the different levels in light of the different priorities. Such a perspective is necessary because of the changing landscape in which healthcare delivery is provided and HIT is used. Early examples of HIT such as the HELP system (Pryor, 1988), the Regenstrief Medical Record System (McDonald et al., 1999), and Brigham Integrated Computing System (Teich et al., 1999) were developed and maintained within the boundaries of individual institutions. The "market" for these systems were contained within the organizations where the systems were implemented. As we move care delivery into the community the market grows and as such creates different categories of coordination between micro and meso levels, as more healthcare delivery is provided via collaborative teams, and between micro, meso, and macro levels as governments demand more accountability for healthcare delivery. These different types of coordination lead to unintended consequences and while HIT is the often criticized as being the direct cause of unintended consequences, in reality it is only the platform. The real source of unintended consequences is the actual implementation of the HIT platform and the coordination provided by HIT.

MULTISIDED FRAMEWORK FOR COORDINATED HEALTHCARE DELIVERY

Figure 8.1 shows our multisided market framework for coordinated healthcare delivery. The framework has three parts to it. First are the three levels of care delivery (micro, meso, and macro). The micro level is defined as individual patients, providers, or organizations. The meso level is the rolling up of several micro-level agents at the collaborative team or organizational level. The macro level is the healthcare system or government level. Second, between each level is a coordination space, which connects it to the other levels. Some elements of coordination occur between adjacent levels, such as micro and meso levels connecting to provide collaborative delivery. Other types of coordination are more distant such as micro level care delivery being used in macro level metrics to measure quality of or access to care. The second part of the framework defines the three levels according to agents, priorities, or needs, and HIT requirements. The third part is HIT that acts as the coordinating platform between the three levels.

Because each level is defined by different agents and priorities or needs, it impacts the HIT support that is needed at each level. These different needs and the HIT

Figure 8.1 Multisided market framework for coordinated healthcare delivery.

support to operationalize the needs often surface as HIT implementation issues. An example is the coordination between the macro level and the other two levels. As healthcare spending costs have escalated governments are wanting to track metrics on healthcare delivery such as patient access or wait times for services (Berwick et al, 2008; Kuziemsky and Peyton, 2016). HIT is useful for monitoring care delivery as it enables tracking of metrics across different settings in a timely manner (Reid et al, 2005). However, monitoring care delivery outcomes can cause a ripple effect through the coordination spaces from the macro to the meso and micro levels due the burden of collecting the additional data. Certain metrics may necessitate collecting additional data and micro-level providers may be faced with workflow issues in collecting the additional data (Kuziemsky and Peyton, 2016). There are also challenges in that data that is collected from front line care delivery may not be of sufficient detail or data standard for macro-level analysis, and that macro-level data may not be an accurate representation of the underlying clinical reality (Martin et al., 2015), which can have further implications as provider reimbursement at the micro level may be linked to meeting certain macro-level performance metrics (Wilson, 2013).

Our multimarket framework provides insight for addressing the above issue by describing how HIT can support coordination across the three levels. At the micro level, individual patients and providers need to be diligent about collecting necessary data about a patient's case. Their needs include access to HIT tools to enable day-to-day management and delivery of patient care. The micro level also bears the burden of responsibility in that it is where patient data collection begins. Poor data collection at this level will ripple through the coordination spaces to the meso and macro levels. However, providers (e.g., physicians) at the micro level are dependent on payments from the macro level and thus there is motivation to provide necessary data to the macro level. HIT needs at the micro level include having access to tools with good usability that do not cause excessive workflow disruptions (Kushniruk et al., 2013). The meso level takes patient care delivery to the team or organizational level. Collaborative care delivery is increasingly becoming the norm for how care is provided. The meso level is dependent on the micro level as issues at the micro level such as an individual provider not completing a requested task will have a ripple effect at the team level. Meso-level HIT needs can include social technologies or tools, such as Web 2.0 tools, in the style of Twitter or Facebook for connecting team members or for coordinating joint decision making. The macro level is where policy and funding begins and has become a bigger player in healthcare delivery due to governments wanting more accountability for healthcare delivery. The macro level enables the micro and meso levels through policy and funding but it also depends on them as no matter how good healthcare policy is it all comes down to how well policy is operationalized at the micro and meso levels. Further, the data the macro level needs in order to analyze and plan healthcare delivery originates from care delivery at the micro and meso levels.

The multimarket framework also emphasizes that achieving cross-level coordination requires trade-offs to be made across all three levels. The different needs at each level can at times be in conflict with each other. These trade-offs are particularly significant in the context of HIT implementation for coordinated care delivery. One example is the connection between the micro and meso levels as part of providing collaborative care delivery. Collaborative-based care delivery at the meso level is conducted by several micro-level agents. HIT has been shown to be valuable for supporting collaboration (Collins et al., 2011; Weir et al., 2011). However, HIT support for collaboration can necessitate trade-offs, e.g., at the micro level, where individual providers may have to endure additional data collection requirements, or complete additional tasks, in the context of working in a collaborative team (Kuziemsky and Bush, 2013).

Overall, our framework emphasizes that a fundamental aspect of HIT implementation for coordinated care delivery is understanding the care delivery and HIT needs at each level, but also understanding the coordination space across levels to determine what trade-offs may need to be made and how to best implement these trade-offs.

CONCLUSIONS

HIT will play a crucial role in the provision of coordinated care delivery. While HIT-mediated coordination is often a good thing it can also be a cause of unintended consequences due to conflicts in the care delivery needs across different levels of care. HIT support for one level of care delivery may cause workflow or data collection issues, at the other levels. To address these issues we need to understand HIT implementation across different levels of care delivery.

This chapter presented a multisided market framework for coordinated healthcare delivery. The framework contains three levels (micro, meso, and macro) of healthcare delivery, a coordination space that connects the three levels, and HIT as the platform for supporting coordination. The key contribution from our framework is defining the coordination space across the three levels. The coordination space helps us understand how each of the three levels both enables and constrains the other levels. Enablement occurs because the three levels depend on each other for the healthcare system to work effectively. Individual care processes at the micro level are the basis for team-based collaborative processes at the meso level, as well as for measuring and monitoring care delivery at the macro level. Subsequently the macro level is where much of the policy and funding originates that enables the micro and meso levels to function by empowering patients and providers. In a well-functioning healthcare system the three levels work in harmony by enabling the other levels.

However, such harmony is often not the norm because each of the three levels has different priorities or needs, and therefore they constrain the other levels. Working in collaborative care delivery at the meso level will often put additional workload on individual providers at the micro level, while macro-level policy intended to innovate or standardize healthcare delivery will cause changes for both micro- and meso-level processes. While the HIT that implements these needs (e.g., additional data collection fields, collaborative care plans) is often blamed for workload or process changes, it is in fact a system level issue due to the multimarket manner in which healthcare is delivered. When the various players in the market have different needs there will inevitably be some degree of conflict. In order to reduce the degree of conflict across the three levels we need to define how the coordination space works across the different levels and also to identify the type and extent of trade-offs that need to be made in order for system level coordination to work. These trade-offs then need to inform the design and evaluation of HIT to support coordinated care delivery.

REFERENCES

Ash, J.S., Sittig, D.F., Poon, E.G.,, et al., 2007. The extent and importance of unintended consequences related to computerized provider order entry. J. Am. Med. Inform. Assoc. 4 (4), 415–423.

Bates, D.W., 2015. Health information technology and care coordination: the next big opportunity for informatics? Yearb. Med. Inform. 10 (1), 11.

Berwick, D.M., Nolan, T.W., Whittington, J., 2008. The triple aim: care, health, and cost. Health. Aff. 27 (3), 759–769.

Blumenthal, D., 2011. Wiring the health system—origins and provisions of a new federal program. N. Engl. J. Med. 365, 2323–2329.

Businger, A., L. Buckel, T. Gandhi, R. Grant, E. Poon, J. Schnipper, et al., 2007. Patient review of selected electronic health record data improves visit experience. *American Medical Informatics Association (AMIA) Annual Symposium proceedings*, p. 887.

Collins, S.A., Bakken, S., Vawdrey, D.K., Coiera, E., Currie, L., 2011. Model development for EHR interdisciplinary information exchange of ICU common goals. Int. J. Med. Inform. 80, e141–e149.

Devlin, A.M., McGee-Lennon, M., O'Donnell, C.A., Bouamrane, M.-M., Agbakoba, R., O'Connor, S., et al., 2016. Delivering digital health and well-being at scale: lessons learned during the implementation of the dallas program in the United Kingdom. J. Am. Med. Inform. Assoc. 23 (1), 48–59.

Hamblin, R., Ganesh, J., 2007. Measure for Measure: Using Outcome Measures to Raise Standards in the NHS. Policy Exchange, London.

Harrison, M.I., Koppel, R., Bar-Lev, S., 2007. Unintended consequences of information technologies in health care: an interactive sociotechnical analysis. J. Am. Med. Inform. Assoc. 14, 542–549.

IOM (Institute of Medicine), 2012. Health IT and Patient Safety: Building Safer Systems for Better Care. The National Academies Press, Washington, DC.

Kuperman, G.J., McGowan, J.J., 2011. Potential unintended consequences of health information exchange. J. Gen. Intern. Med. 28 (12), 1663–1666.

Kushniruk, A., Nohr, C., Jensen, S., Borycki, E.M., 2013. From usability testing to clinical simulations: bringing context into the design and evaluation of usable and safe health information technologies. Contribution of the IMIA Human Factors Engineering for Healthcare Informatics Working Group. Yearb. Med. Inform. 8, 78–85.

Kuziemsky, C.E., 2015. A review of social and organizational issues in health information technology. Healthc. Inform. Res. 21 (3), 152–161.

Kuziemsky, C., 2016. Decision-making in healthcare as a complex adaptive system. Healthcare Manage. Forum 29 (1), 4–7.

Kuziemsky, C.E., Peyton, L.A., 2016. A framework for understanding process interoperability and health information technology. Health Policy Technol. 5 (2), 196–203.

Lusignan, S.L., Mold, F., Sheikh, A., Majeed, A., Wyatt, J., Quinn, T., et al., 2014. Patients' online access to their electronic health records and linked online services: a systematic interpretative review. BMJ Open 4, e006021.

Martin, G.P., McKee, L., Dixon-Woods, M., 2015. Beyond metrics? Utilizing "soft intelligence" for healthcare quality and safety. Soc. Sci. Med. 142, 19–26.

McDonald, C.J., Overhage, J.M., Tierney, W.M., Dexter, P.R., Martin, D.K., Suico, J.G., et al., 1999. The Regenstrief medical record system: a quarter century experience. Int. J. Med. Inform. 54 (3), 225–253.

McGlynn, E.A., Shekelle, P., Hussey, P., 2008. Developing, Disseminating and Assessing standards in the National Health Service. RAND Health, Cambridge.

Mennemeyer, S.T., Menachemi, N., Rahurkar, S., Ford, E.W., 2016. Impact of the HITECH act on physicians' adoption of electronic health records. J. Am. Med. Inform. Assoc. 23 (2), 375–379.

Mettler, T., Eurich, M., 2012. What is the business model behind e-health? A pattern based approach to sustainable profit. ECIS 2012 Proceedings. Paper 61.

O'Malley, A.S., Grossman, J.M., Cohen, G.R., Kemper, N.M., Pham, H.H., 2010. Are electronic medical records helpful for care coordination? Experiences of physician practices. J. Gen. Intern. Med. 25 (3), 177–185.

Pryor, T.A., 1988. The HELP medical record system. MD Comput. 5 (5), 22–33.

Reid, P.R., Compton, W.D., Grossman, J.H., Fanjiang, G., 2005. Building a Better Delivery System. A New Engineering/Health Care Partnership. The National Academies Press, Washington, D.C.

Rochet, J.-C., Tirole, J., 2003. Platform competition in two-sided markets. J. Eur. Econ. Assoc. 1 (4), 990–1029.

Rodriguez, E.S., Thom, B., Schneider, S.M., 2011. Nurse and physician perspectives on patients with cancer having online access to their laboratory results. Oncol. Nurs. Forum. 38 (4), 476–482.

Samal, L., Dykes, P.C., Greenberg, J., Hasan, O., Venkatesh, A.K., Volk, L.A., et al., 2013. The current capabilities of health information technology to support care transitions. AMIA. Annu. Symp. Proc., 1231.

Teich, J.M., Glaser, J.P., Beckley, R.F., Aranow, M., Bates, D.W., Kuperman, G.J., et al., 1999. The Brigham integrated computing system (BICS): advanced clinical systems in an academic hospital environment. Int. J. Med. Inform. 54 (3), 197–208.

Vezyridis, P., Timmons, S., 2014. National targets, process transformation and local consequences in an NHS emergency department (ED): a qualitative study. BMC. Emerg. Med. 14 (1) art. no. 12.

Weir, C.R., Hammond, K.W., Embi, P.J., Efthimiadis, E.N., et al., 2011. An exploration of the impact of computerized patient documentation on clinical collaboration. Int. J. Med. Inform. 80, e62–e71.

Wilson, K.J., 2013. Pay-for-performance in health care: what can we learn from international experience. Qual. Manage. Health. Care. 22 (1), 2–15.

Business Models

PART V

Business Models

CHAPTER 9

Explaining Healthcare as a Two-Sided Market Using Design Patterns for IT-Business Models

M. Eurich[1,2] and T. Mettler[3]
[1]ETH Zurich, Zurich, Switzerland
[2]University of Cambridge, Cambridge, United Kingdom
[3]University of Lausanne, Lausanne, Switzerland

INTRODUCTION

With the advent of the Internet the disruptive impact of the "Information Revolution" was unleashed, which thoroughly changed markets, economies, and industry structures. The Internet became a major distribution channel for goods and services and in the course of the digitization wave the impact of e-commerce spread out to changing products, services, consumer behavior, and eventually jobs and labor markets (Drucker, 1999; Freeman and Louçã, 2001). The Information Revolution poses a challenge to designing viable business models for enterprises in almost all industries, but particularly for IT-reliant services, as their design is strongly interrelated with the business model (Al-Debei and Avison, 2008; Drucker, 1999; Lindgardt et al., 2009).

In the extant literature a variety of different connotations and meanings are assigned to the term "business model" (Burkhart et al., 2011). The first part—business—is attributed to the conduct of commercial transactions in order to accomplish a company's mission and the second part—model—to a simplified picture of the reality that consists of elements and their relations to each other. In this article, we refer to a *business model* as a theoretical concept that reflects the way a company conducts commercial transactions, and thereby creates and captures value (Baden-Fuller and Morgan, 2010; Zott and Amit, 2010).

Baden-Fuller and Haefliger (2013) perceive the role of the business model as mediating between technological innovation and firm performance and Chesbrough (2010) emphasizes that the same technology commercialized in different ways may result in different economic outcomes. In competitive markets, innovation of the business model is considered a means to gain a competitive edge and to (pro) actively enact changes on the market (Lindgardt et al., 2009; Pohle and Chapman, 2006). Accordingly, *business model innovation* refers to the creation or reinvention, and thus to the introduction of something new or different in doing business. Business model

153

innovation is increasingly recognized as a promising way to generate competitive advantage especially in rapidly changing environments driven by new technologies or new regulations (Chesbrough, 2010; Sako, 2012; Teece, 2010) and has become a subject of innovation in itself, because firms can compete through their business models (Casadesus-Masanell and Zhu, 2013). However, neither technological nor business-enabled innovation may lead to economic success without their respective counterparts. Many scholars point out the need for a coevolutionary approach that combines IT-driven value creation with the entrepreneurial perspective on innovation (Davidson et al., 2012; Giudice and Straub, 2011; Hackney et al., 2004; Kagermann et al., 2010; von Krogh and Spaeth, 2007).

Designing and implementing even a loosely coupled system is not a trivial undertaking, and the strong interdependence between the IT-artifact and the business model makes it even harder for managers, entrepreneurs, and IT strategists to master this task. Prior research found a variety of reasons why organizations struggle with business model innovation, such as issues related to novel and changing technologies, uncertain assumptions about the environment of the organization, varying worldviews within the firm, or a lack of expertise and negligence in the act of "designing" a business model (e.g., Pohle and Chapman, 2006; see also Drucker, 1984).

Despite the fact that there has been a recent increase in research interest on business models (Klang et al., 2014), emphasis on studying "how" to design business models has been rather little as compared to investigating different connotations and meanings of "what" comprises a business model (Osterwalder and Pigneur, 2013). Thus far, the mechanisms of how the inherent logic of business model works are typically not analyzed in detail. To address this shortcoming, we derive the concept of design patterns from architecture (Alexander et al., 1977) and software engineering (Gamma et al., 1995) and apply it to business model design. Analogously to the understanding of design patterns in other disciplines, a business model design pattern (BMDP) can be perceived as a formal means of documenting the generic logic behind a business model for a particular context or problem situation.

In this chapter, we introduce the concept, requirements, and usefulness of BMDP first generically before we emphasize the characteristics of the two-sided market BMDP in healthcare as a prime example. Two-sided markets can be understood as an elaboration of the concept of network effects (Katz and Shapiro, 1985; Shapiro and Varian, 1999). As BMDP, two-sided markets are characterized by an economic platform that features two (or more) distinct user groups, who depend on each other. The platform owner provides each user group with network benefits and creates value primarily by enabling interactions between two or multiple distinct affiliated customers (Eisenmann et al., 2006; Parker and Van Alstyne, 2005). Two-sided markets can be found in many industries, such as credit cards (cardholders and merchants), real estate brokerage (renters and landlords), Internet portals, and search engines (sites, surfers), magazines (readers and writers), yellow page directories (readers and businesses), and is also considerably prominent in healthcare.

To begin with, we propose a generic approach that abstracts the basic idea of a business model from its organizational environment. The abstraction of the business model idea can be generalized and therefore be transferred and reused to commercialize another service or product, such as we will illustrate with e-health services.

A well-designed business model can be a competitive advantage (Lindgardt et al., 2009; Pohle and Chapman, 2006). First mover advantages can support an organization in gaining a competitive edge (Lieberman and Montgomery, 1988). However, the basic mechanisms of the business model can often not be protected, because important parts of the business model such as the value proposition and the pricing will be identified by the consumers and business analysts. Therefore, the abstraction of a successful business model could be reused in another situation.

In fact, many of the business model innovations are not as innovative as they may appear in the first place. Much of what is labeled "innovation" is not really a new invention, but rather the adaptation of an established model to a new domain, or the combination and integration of elements of successful existent models into a new one. Technology is one of the major enablers in mobilizing business models (Baden-Fuller and Haefliger, 2013; Bharadwaj et al., 2013; Gregor et al., 2006). For example, Amazon successfully mobilized Sears Roebuck's business model of selling books via a catalog to the Internet (Baden-Fuller and Haefliger, 2013); 1995-founded EasyJet applied the low-cost airline carrier business model that Southwest Airlines already introduced in the 1970s (Hunter, 2006; Slywotzky, 1996, p. 117 et seq.; Yip, 2004); Apple adapted Gillette's razor-and-blades business model to iTunes (i.e., the "blades") in combination with the iPod (i.e., the "razor") (Johnson et al., 2008); and Hilti excelled in applying a servitization approach for high-end power tools for the construction industry rather than in inventing a new business model from scratch (Enkel and Mezger, 2013; Johnson et al., 2008; Neely, 2008).

In this context it is also important to clearly separate between business model and strategy: A business model is part of the strategy, but it is not congruent with the strategy (Baden-Fuller and Haefliger, 2013; Casadesus-Masanell and Ricart, 2010; Teece, 2010). A business model can be considered as intermediary between strategy and business processes (Al-Debei and Avison, 2008; Casadesus-Masanell and Ricart, 2010). Different business models could be equifinal, i.e., they could realize a certain strategy in an equally acceptable manner.

A BMDP can provide a business model designer or information system developer with some inspirations and guidelines, but only its successful adaption, adjustment, and potential combination with another BMDP can serve as a starting point to design a viable business model that finds its reflection in IS design. The purpose, key principles, and contributions of BMDPs are explained as follows in this chapter: we discuss the generic idea of design patterns and describe four basic principles, which need to be adhered to when defining a pattern. We then present how to identify, sketch, and substantiate a BMDP and use it for describing the "two-sided market" business model of healthcare. We conclude with a discussion of the practical and theoretical implications of our proposition and give recommendations for future research.

GENERIC META-REQUIREMENTS OF BUSINESS MODEL DESIGN PATTERNS

Managers', entrepreneurs', or IT strategists' burden to innovate the business model of their enterprise can often be transferred into the task of adapting an established model to their specified requirements or latent needs. Frequently what they need for this task is an "exemplar" (Baden-Fuller and Morgan, 2010), a "recipe" (Sabatier et al., 2010), or a "template" (Baden-Fuller and Winter, 2005; Doganova and Eyquem-Renault, 2009). To guide them through the conceptualization phase of a business model, such a "template" needs to fulfill the following meta-requirements:

- *Focus on the logic inherent in a business model:* A business model is a model, and it should become clear how and why it works. By focusing on the logic inherent in a business model, critical mechanisms for business success should become evident and generalizable across firms, and transferable from time and space (Al-Debei and Avison, 2010; Baden-Fuller and Morgan, 2010; Gordijn and Akkermans, 2001; Zott and Amit, 2010; Zott et al., 2011).
- *Clearly separate between business model and strategy:* In order to avoid confusion with strategic objectives that can be unique to a particular organization, the mechanisms of a business model should be clearly separated from strategy (Baden-Fuller and Haefliger, 2013; Casadesus-Masanell and Ricart, 2010; Teece, 2010). BMDP aims at a generalization of the basic business model idea, which should not be confused with an organization's strategic assets and competitive advantages.
- *Include a graphical representation:* A graphical representation can contribute much to the clarification of how and why a business model works. Assumptions, decisions, and consequences become visible. A graphical representation is supportive in capturing, discussing, and reusing business models (Gordijn and Akkermans, 2003; Veit et al., 2014; Zott et al., 2011).
- *(Re)use business model language elements:* To make the basics of a business model replicable and to explain how value is created, the structure and entities of a business model and their relations should be represented (Al-Debei and Avison, 2008; Burkhart et al., 2012; Zott et al., 2011). It is indispensable to deploy a clearly

defined set of language elements and reuse them in as many business models as possible throughout and across the company. A widespread and constant use and further refinement of such a vocabulary may possibly lead to language elements with so-called reference character (vom Brocke, 2007).

In line with discussions on business model design and re-use—or at least getting inspirations for the design (Pateli and Giaglis, 2004)—different options for capturing the exemplary business model knowledge are proposed:

- *Cases* give business model designers valuable insights into decisions and the specific circumstances that made the company agree on the particular chosen business model design (Girotra and Netessine, 2011; Sosna et al., 2010; Yunus et al., 2010). In the business model literature cases usually describe the development and success factors of a concrete organization's business model. Typically, business model case studies incorporate the specific requirements of the focal organization and, therefore, they often do not clearly separate the business model from the strategy. A graphical representation is often missing.

- *Taxonomies* provide business model designers a description of different types of business models and the reuse of existing analytical work (Rappa, 2004; Timmers, 1998). Although they do not merge the business model types with strategy, taxonomies tend rather to provide a generic description of the types in a particular field (e.g., electronic markets or high performance computing centers in higher education) than to explain the inherent logic that makes the business model types work. Graphical representations are typically not part of taxonomies.

- *Component-based approaches* provide the designer with a predefined set of constituent components which, combined and instantiated, build a concrete business model (Afuah and Tucci, 2001; Chesbrough and Rosenbloom, 2002; Morris et al., 2005). Following a bottom-up approach, the components are typically instantiated first before they get combined with each other. As a consequence thereof the connections between the components and the mechanisms of how the business model works emerge over the design process. However, there is no particular focus on the logic inherent in the business model in the first place. Component-based approaches present the constituent building blocks, but typically do not feature a graphical representation of the logic inherent in the business model. Strictly speaking, component-based approaches do not provide explanations of the business logic. They instantiate a rather anatomic way of the business model that addresses the entities of a business model, but not the structure and relations within that explain how value is created.

- *Conceptual models* go beyond the component-based approaches by addressing interrelations and interdependencies between the components (e.g., Al-Debei and Avison, 2010; Boutellier et al., 2010; Johnson et al., 2008; Osterwalder and Pigneur,

Table 9.1 Fulfillment of meta-requirements by existing techniques to business model innovation (solution space)

Solution space	Meta-requirements			
	Focus on logic inherent in the business model	Separation from business strategy	Graphical representation	Reuse of language elements
Cases	✗	✗	✗	✗
Taxonomies	✗	✓	✗	✓
Components	✗	(✓/✗)[a]	✗	(✓/✗)[a]
Conceptual	(✓/✗)[a]	(✓/✗)[a]	(✓/✗)[a]	(✓/✗)[a]
Causal loops	✓	✗	✓	✗

[a]*Note*: Depends on the particular author.

2002; Rai and Tang, 2014; Samavi et al., 2009). In general, they feature a reasonable morphological and some of them also a graphical representation. There is, however, a fatal tendency to sacrifice a clear focus on the logic inherent in the business model for model-internal consistency, particularly when the logic of the business model cannot exactly be expressed with the conceptual framework.

- *Causal loop diagrams* (system dynamics) focus on the mechanisms of a business model and describe its underlying logic (e.g., Casadesus-Masanell and Ricart, 2010, 2011; Kiani et al., 2009; Seelos and Mair, 2007). They often rely on decisions, choices, and their consequences (cf. e.g., Casadesus-Masanell and Ricart, 2011), and as a consequence causal loop diagrams tend to merge business model aspects with strategic objectives and a morphological representation appears rather indirectly and not explicitly.

Table 9.1 summarizes to which extent existing business model design techniques (solution space) meet business model design requirements that allow for transferability and generalizability.

Our goal is to complement these techniques that assist the designers in the business model design process with an approach that particularly focuses on the "logic inherent in a business model" and "graphical representations" (Veit et al., 2014). The aim is to suggest a template to transfer and mobilize business model logics and make them transferable and generalizable across organizations.

To this end, we propose BMDPs as formal means of documenting the generic logic behind a business model. In the following section, we outline the antecedents of BMDPs, from which we derive generic principles. In the section that follows, we clarify the use of BMDP in practice and explain what is specific to two-sided markets in healthcare and why it is important to deliberate about how to attain sustainable profits in combination with a well-defined value proposition for both health professionals and patients.

GENERIC PRINCIPLES OF BUSINESS MODEL DESIGN PATTERNS

The notion of *design patterns* was first introduced in the late 1970s in the field of architecture, specifically in urban planning, where it was used to capture collective design knowledge. Presuming that the actual inhabitants know more about the buildings they need than any architect is able to envisage, Alexander et al. (1977) formulated a so-called *pattern language* that should enable both architects and nonarchitects to externalize proven practical, safe, and attractive designs. In the 1990s, design patterns became extensively popular in software engineering, where they refer to archetypal solutions for recurring programming problems (Buschmann et al., 2007; Gamma et al., 1995).

Albeit the fact that design patterns for business models have already been discussed in literature (Abdelkafi et al., 2013; Casadesus-Masanell and Ricart, 2011; Frankenberger et al., 2013; Mackenzie et al., 2014; Mettler and Eurich, 2012; Osterwalder and Pigneur, 2010), a shared conceptualization is still missing and a clarification of the meaning is required.

In order to provide such a clarification and taking upon the prior definition of business models, a BMDP can be understood as an *abstract illustration of an instantiated business model in practice*. In this sense, a BMDP can be perceived as formal means of documenting the generic logic behind a business model for a particular context or problem situation. In line with previous experiences from architecture (Alexander et al., 1977) and software engineering (Buschmann et al., 2007; Gamma et al., 1995), we argue that a BMDP must adhere to the following principles (cf. Table 9.2).

Table 9.2 Generic principles of a business model design pattern

Principle	Description
Purpose and scope (*causa finalis*)	A description of one or more situations or contexts in which the design pattern is applicable as well as the goal that can be achieved with this business model.
Entities of business model (*causa materialis*)	Representation of the entities of interest in the business model. At the minimum this consists of a description of the most relevant actors, value activities (i.e., operational activities yielding a profit or utility increase assigned to a specific actor), and value objects (i.e., an object that is of value for one or more actors).
Functioning and blueprint of business model (*causa formalis*)	Visual illustration of the abstract architecture of the business model and the logic behind it, respectively the relationships between the main entities of the business model.
Exemplary instantiation (*causa efficiens*)	References to exemplary firms and industries that use the design pattern or literature that further describes the design pattern.

Principle 1: *A design pattern should be capable of solving one or more generic problems and be applicable in one or more contexts.*

In designing a business model, this is the *causa finalis* (Falcon, 2011), i.e., the description of one or more situations or contexts in which the business model is applicable as well as the goal that can be achieved with this business model. For instance, the goal of a business model for an Internet start-up could be to commercialize a software system within a specific market segment (Chesbrough and Rosenbloom, 2002).

Principle 2: *A design pattern should describe the major entities of interest in a way that is comprehensible for entrepreneurs.*

In business model design this refers to the *causa materials* (Falcon, 2011), i.e., to the representation of the different entities of interest in the business model, like for example key actors, value activities, value offerings, value exchange, or market segments (Al-Debei and Avison, 2010; Burkhart et al., 2012; Gordijn and Akkermans, 2001; Hedman and Kalling, 2003; Mettler, 2013).

Principle 3: *A design pattern should provide insights into the form and function of a design.*

Related to business models, this refers to the *causa formalis* (Falcon, 2011), i.e., an abstract "blueprint" of a business model and its function (Gregor and Jones, 2007). This means a visual illustration of the abstract architecture of a business model, including the relationships between the identified entities and the logic behind it. Different representational forms are conceivable (Gordijn and Akkermans, 2003; Osterwalder, 2004; Peinel et al., 2010; Penker and Eriksson, 2000; Pynnonen et al., 2008; Samavi et al., 2009; Yu et al., 2011).

Principle 4: *A design pattern should be based on the foundation of a real design.*

According to Gregor and Jones (2007) this refers to the *causa efficiens*, which means the mode of implementation, i.e., a "description of processes for implementing the theory (either product or method) in specific context" (p. 322). On the context of business models, this could mean an explanation of how a business model can be implemented (Benlian et al., 2009; Iacob et al., 2012; Sosna et al., 2010) and a description of exemplary cases of firms or industries that have adopted this particular kind of business model (Burkard et al., 2012; Chesbrough and Rosenbloom, 2002; Kinder, 2002; Schief, 2014).

BUSINESS MODEL DESIGN PATTERNS IN PRACTICE

Despite the fact that no common approach for documenting business model-related patterns or "classes" of business models exists, we found a variety of exemplary BMDP, implicitly outlined in the extant literature (cf. Table 9.2). For instance, Picker (2010) describes the entities of interest and the mechanism of the so-called razors and blades pattern, but no visualization of the mechanism is provided. Osterwalder and Pigneur (2010, p. 54 et seq.) suggest a visualization and demonstrate its usefulness by using several patterns as examples, like the "long tail" or "inverted freemium." They also provide several instantiations, i.e., companies that have instantiated these identified patterns. Yet, the scope is not explicitly described, i.e., the particular situation or context in which the BMDP could be applied. Bughin et al. (2010) provide a couple of instantiations of the BMDPs "freemium" and "multisided," but there are no detailed insights on how the mechanisms work and there is no visualization.

In order to ease the documentation of BMDP and to facilitate their proliferation in research and practice, it is important to provide a systematic way of formalizing existing as well as creating new design patterns. The specification of such a BMDP is partially an act of practical problem-solving, which is frequently adhered to constructive (Lukka, 2003) or design-oriented research methodologies (March and Smith, 1995). In line with Archer (1969) the logical nature of the act of practical problem-solving must not necessarily be dependent on the solution that is being developed. Accordingly, several researchers have tried to systematize a general design process by decomposing and classifying the central design activities and by finding rationales for effective problem-solving (e.g., Coyne, 1988; Nunamaker et al., 1990; Takeda et al., 1990). Although a commonly agreed upon procedure model is lacking, three mutually dependent yet methodologically distinct main phases can be identified in all problem-solving approaches: (1) problem analysis, (2) design, and (3) evaluation.

In the context of BMDP, problem analysis may refer to the identification of undocumented or partially documented business logics. Design can allude to the act of formally describing and figuratively illustrating the identified generic business logic. Finally, evaluation may be understood as the provisioning of underlying justificatory knowledge (e.g., literature sources, exemplary real-life cases) that such a business logic de facto can be observed in practice (Table 9.3).

EXPLAINING HEALTHCARE AS TWO-SIDED MARKET

To explain how the BMDP concept works, we will use the *two-sided market* pattern as an example for explaining how the healthcare domain frequently works. The starting point, according to Järvinen (2007), is a sound analysis of the domain or

Table 9.3 List of exemplary business model design patterns documented in literature

Pattern name	Short description	Exemplary instantiations	Exemplary reference
Freemium	Basic services are offered for free, while a premium is charged for an advanced service	VoIP Services (e.g., Skype)	Anderson (2009)
Two-sided or multisided	Value creation is based on the interaction among the parties	Healthcare	Bughin et al. (2010), Eisenmann et al. (2006)
Shareholder model	A group of service consumers who need a particular service pool their money in order to finance the required product. The service provider takes care of the acquisition, maintenance, and operation	High Performance Computing or Mobility services (e.g., ZipCar)	Keegan (2009), Teece (2010)
Crowd sourcing/open	The commitment of motivated individuals produce value for the organization for free	Open source journalism (e.g., *Huffington Post*)	Chesbrough (2010)
Inverted freemium	Customers pay a premium in order to be entitled to free service	Insurance	Osterwalder and Pigneur (2010)
Razors and blades	A customer is lured with a low initial investment and by relying on lock-in effects, profit is made from the sales of complementary goods	Computer hardware and accessories (e.g., printers and ink cartridges)	Picker (2010)
The long tail	On the basis of a solid platform, aggregated sales of numerous niche products are provided	Print-on-demand and self-publishing (e.g., Lulu.com)	Osterwalder and Pigneur (2010)
Affiliate programs	A service is provided for free or almost for free. The service provider receives a revenue share for referring customers to one or more affiliates	Voucher services (e.g., The Bargainist, Money Saving Expert)	Wooldridge and Schneider (2010)
As a service	Only the consumption of a service is charged, not the product itself	On demand software (e.g., Salesforce, Netsuite)	Benlian and Hess (2011)

"innovation opportunities" as an essential first step for clarifying relevance and orientation. A better understanding of domain-relevant business logics can, for instance, be gathered from the relevant literature, empirical observations, or personal involvement in practice. Key questions that may support the identification of a generic business logic are:

- *What is the particular situation or context?* In our illustrative example, the two-sided market BMDP can refer to situations in which one party has a considerable interest that another party is subsidized or given easier access to a service, e.g., insurance companies that want their policyholders to track their health status in a personal healthcare record, or hospitals that are interested in having an electronic medical report (e.g., including a patient history, diagnostic findings) from all of their referring doctors. Typically the service to be delivered is either very complex from a technical point of view or there is a lack of trust when it is provided by the interested party on its own. Besides the role of the service consumer and the service provider, there is also the need for a mediator or broker who has a certain level of trust or the required competencies to bundle and operate these e-health services.

- *Which actors play an important role in this particular context?* In the simplest form, there are only three key actors in the two-sided BMDP: a service provider (or focal company), a service consumer, and a broker. This broker has a pivotal role within this business logic. Being a kind of "platform" for all sorts of services, the broker has the crucial task of balancing the subsidies and revenues from multiple parties in order to ensure both a large customer base (who trust the broker) as well as a high quantity and quality of services. The former especially is difficult to achieve because there is a high risk that the broker is misled by short-term profits which, e.g., may result in not investing enough in security or marketing (Dunbrack, 2011).

- *What specific problems do the identified actors have?* It is frequent that brokers are confronted with a so-called chicken and egg problem, i.e., members of each group (service providers and service consumers) are willing to participate in the platform only if they expect many members from the other side to participate. In addition, there might be the problem that there is no common standard such that service providers' offerings can easily be integrated onto their platform (i.e., each service provider develops e-health services with their own proprietary standard) or classifications that help service consumers to find the right offerings. Further, the market of e-health service users can be extremely fragmented, which increases the complexity of providing customized or patient-centric services. Finally, additional complexity is added, if the broker operates in several countries with very different market structures and regulations.

- *What are the strategic objectives?* As the name suggests, two-sided markets can be differentiated from a sellers' and a buyers' perspective. In this sense, a strategic objective could be to provide and maintain a platform that contains all necessary e-health services for patients, while is attractive enough for service providers to showcase their particular offerings. A primer objective should be to identify mechanisms that create lock-in effects on both sides, as the value of joining the platform depends on expectations about the opposite network size. Accordingly, a balance between the interests of service providers and service consumers is of utmost importance for two-sided markets.

Once having gathered relevant information related to purpose and scope (causa finalis) as well as main entities of the business model (causa materialis), the next step should be dedicated to describing the functioning and blueprint of the business model (causa formalis). As referred to in the literature (McGrath, 2010; Osterwalder and Pigneur, 2010; Timmers, 1998), a key question that needs to be answered is: *How can value be created, delivered, and captured under consideration of the domain and other contextual factors?*

For answering this question, visual designs or symbolic language could be used. It is important to not only provide narratives in the form of textual descriptions but also by a graphical blueprint of the BMDPs as well. As described before, there are several representational forms of documenting a business model (cp. Kundisch et al., 2012). It is important to note that no specific modeling language for BMDP does still exist.

To illustrate the value flows in the two-sided market BMDP, we use *e3-value* (as proposed by Gordijn and Akkermans, 2001; Veit et al., 2014), because it is a lightweight yet expressive notation for capturing the logic of the value exchange between the focal enterprise and the service consumers. It is also conceivable to use drawings, or other graphical representation means. The advantage of using a well-known modeling language is the broad comprehensibility by experts. In this sense, also more general modeling languages such as UML, EXPRESS-G, and the like can be used (Gordijn and Akkermans, 2001; Veit et al., 2014).

For specifying the blueprint of the two-sided example, we only used basic modeling elements: the main actors (*rectangles*), the market segments (*stacked actor rectangles*), the value activities (*rounded rectangles*), and finally the value exchange (*lines between two interfaces, whereas the annotation represents the exchanged value object*). The corresponding blueprint is illustrated in Table 9.4. It is further important to state that a graphical representation of the BDMP allows both experts and rookies in business model designs to quickly catch sight of the business logic behind the pattern.

After having delineated the business logic behind the design pattern, a crucial step is to provide some evidence of its working as well as some exemplary instantiations (causa efficiens) from which entrepreneurs can be inspired. The evaluation of a BDMP should include, according to Balci (1998), both a *verification* of the proper use of the notation used for illustrating the pattern (was the BDMP designed in the right manner?) and a *validation* of its truthfulness and applicability for the intended context (does

Table 9.4 Lightweight template for documenting a business model design pattern
Description of two-sided market BMDP

Purpose and scope	Two-sided market refers to a situation where two distinct parties interact with each other through a common platform (broker). Typically, one party has a considerable interest that another party is subsidized or given easier access to a service, e.g., insurance companies that want their policyholders to track their health status in a personal healthcare record, or hospitals that are interested in having an electronic medical report (e.g., including a patient history, diagnostic findings) from all of their referring doctors. The value of joining the platform depends on expectations about the opposite network size.
Main entities	A two-sided market is composed of service consumers (end-users), service providers, and a broker.
Functioning and blueprint	In order to improve transaction costs, the broker operates and maintains a platform where the different service consumers and service providers are matched. In doing so, the broker needs to create added-value on both sides; for example by providing service providers with detailed information about services consumers on their platform, or by bundling patient-specific e-health services to suffice particular needs of service consumers. 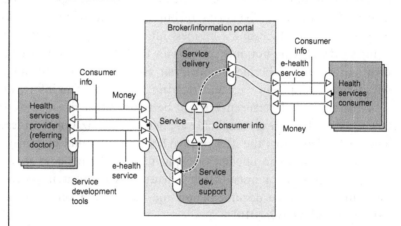
Exemplary instantiations	Examples of two-sided markets can be found in various industries such as credit cards (cardholders and merchants), real estate brokerage (renters and landlords), Internet portals and search engines (sites, surfers), magazines (readers and writers), yellow page directories (readers and businesses), shopping malls (shoppers and shops), publication software (authors, readers), or videogame consoles (gamers and developers).

the BDMP reflect reality in the right manner?). The latter can be demonstrated by providing references to companies that have implemented this pattern.

CONCLUSION

In this chapter, we propose to apply the concept of design patterns to the domain of business model innovation. We introduce a generic template to systematically document business model design patterns by describing (1) their purpose and scope, (2) the core entities involved, (3) the functioning and blueprint of the business model, and (4) examples of businesses that adopted this pattern in practice.

The design of a business model influences the realization of strategic objectives (Osterwalder and Pigneur, 2013). Casadesus-Masanell and Ricart (2010) found that "strategy refers to the choice of business model through which the firm will compete in the marketplace" (p. 196). The application of a BMDP can help to achieve the strategic objectives and can even increase a company's competitiveness. In case there is no BMDP that matches the strategic goals and situation of a company, the constructs of other BMDP may simply be an inspirational source for entrepreneurs to be used for communicating business model ideas with others. Especially the thus far widely neglected visualization (blueprint of business model, causa formalis) can be helpful in illustrating and discussing a business model during the IS development process. In doing so, BMDPs are not only relevant for the discussion concerning dedicated adjustments of the future business model, but also crucial for informing design decisions within the development stage and/or adaptation stage of an information system.

From a practical perspective it can be noticed that the idea of design patterns for business models is already implicitly applied by organizations—at least to some extent. There is evidence that the idea of BMDPs is in the mind of managers, entrepreneurs, and IT strategists in an implicit or explicit form (Brambilla et al., 2012; Morris et al., 2005). The process of instantiating a pattern is of great importance in practice in order to structure the documentation and communication of business model designs (Al-Debei and Avison, 2010).

In addition, it ensures that all of the important elements of its design always remain in sight. For instance, determining all of the components of a business model (like in research on components) in a self-consistent manner is not sufficient if one loses sight of the scope and purpose of the business model. Furthermore, BMDPs open the "black box" of business model design and describe how business models capture value for the customers of an enterprise and for the enterprise itself. This could be of great value for entrepreneurs: they are not only provided with a stereotypical business model, but they also get a visualization of the abstract logic of business models, their processes and relationships among their entities. Finally, and perhaps as the largest benefit for entrepreneurs, BMDPs present a standardized way to communicate, present, and discuss business model designs.

However, the BMDP approach also has some weaknesses and presents some challenges: the approach is no silver bullet that guarantees a useful design of a business model. There may always be a solution that suits the current requirements better than an existing design pattern. Using an inappropriate design pattern leads to a poor outcome, and in terms of business models, this can relate to low profit margins or even financial losses and unsatisfied consumers. Inexperienced companies may trust standard design patterns too much and may run the risk of randomly combining design patterns if they mistakenly assume that mixing several proven design elements will result in a good design (Pateli and Giaglis, 2004). For decision makers it is essential to test the assumptions that underlie the application of one or more BMDPs. When designing a business model, they should thoroughly think about why it should work and what the relevant parameters are which cause changes in the system, especially in volatile or changing environments, such as healthcare. What was yesterday believed to be an appropriate design (e.g., expert systems instead of patient-centered systems) and sustainable business practices (selling patient data to third parties instead of privacy-aware and transparent information policies) may not be viable today. A business model is permanently subject to improvement and therefore decision makers must constantly question the assumptions that underlie their business models (Gamma et al., 1995; Pree, 1994). That is not at all to say that a business model should always be changed. Indeed, a well-chosen business model design is robust and able to outlive market fluctuations and may even have the power to enact changes upon the market (Drucker, 1994). However, it is important to understand the time when a particular assumption about the market is no longer true.

Moreover, a BMDP is no replacement for a missing corporate strategy. Since a BDMP is tailored to support specific strategic objectives, fundamental considerations as to branding, commercialization, market segments, etc. have to exist beforehand. A BMDP needs to be instantiated! Each company needs to create the basic design in accordance with its business. Adapting a BMDP to the context and matching it with the capabilities of the company is required in order to gain a competitive advantage. A design pattern describes a design in principal and its prototypical application, but each company needs to create the basic design in accordance to its business. For instance, many e-health service providers base their business models on the same BMDP (two-sided market), but their concrete manner of operation is different. A strategic advantage can only emerge through this last step, since BMDPs are generic and applicable to many different settings. It is the same in architecture: a room needs a door and at least one window, but there a many different ways of designing a room, placing the door and windows, and realizing the beauty of the room.

Finally, the truth content of a BMDP and the applicability of a BMDP to an actual business model are hard to evaluate. A BMDP must be internally coherent, i.e., the constructs that build the BMPD must be well-matched, for example the entities of interest (causa materialis) must be connected via value and revenue streams and visualized (causa formalis) in a way that the business model has the potential to fulfill its purpose, and within the realms of possibility (causa finalis). However, the practical application of a BMDP is problematic: the issue is how to measure the different dimensions like truthfulness, relevance, and success. There seems to be no way to judge this. Again, we draw an analogy to design patterns in architecture: a room should have at least one window and it should have a door to enter. But how can you test the truthfulness as long as there may be other options of illuminating a room and entering it? For BMDPs, one could argue that a BMDP fulfills its conditions of existence when businesses that apply the pattern accomplish a specific mission with it (Chesbrough, 2010). However, a BMDP's impact on a company's success of completing a mission is indistinguishable from several confounding factors and the accomplishment cannot easily be ascribed to the decision of applying a BMDP. It depends very much on how the basic BMDP is realized and adapted to the specific business needs and requirements. Up to now, the power of BMDP is also—to some extent—restricted to the expressiveness of the visualization language. A business model designer cannot fully express his design ideas if the language does not provide him with all necessary constructs. While e3-value, i*, UML, EXPRESS-G, and the like may qualify for being used to draw a BMDP blueprint, they were originally not designed for that purpose and hence they may limit the verve of BMDP. Finally, there is a need to find ways to improve the evaluation procedure of BMDP. However, the same way that the pattern concept revolutionized the way in which buildings or software is designed, we believe that the presented approach will have similar positive effects on the design of healthcare business models.

REFERENCES

Abdelkafi, N., Makhotin, S., Posselt, T., 2013. Business model innovations for electric mobility — what can be learned from existing business model patterns? Int. J. Innov. Manage. 17 (1), 1340003.

Afuah, A., Tucci, C., 2001. Internet Business Models and Strategies: Text and Cases. McGraw-Hill Higher Education, New York.

Al-Debei, M.M., Avison, D., 2008. Defining the business model in the new world of digital business. In: Proceedings of the 14th Americas Conference on Information Systems (AMCIS), Toronto, Canada, pp. 1–11.

Al-Debei, M.M., Avison, D., 2010. Developing a unified framework of the business model concept. Eur. J. Inf. Syst. 19 (3), 359–376.

Alexander, C., Ishikawa, S., Silverstein, M., 1977. A Pattern Language: Towns, Buildings, Construction. Oxford University Press, New York.

Anderson, C., 2009. Free: The Future of a Radical Price: The Economics of Abundance and Why Zero Pricing is Changing the Face of Business. Random House, New York and London.

Archer, L.B., 1969. The structure of the design process. In: Broadbent, G., Ward, A. (Eds.), Design Method in Architecture. Lund Humphries, London, pp. 76–102.

Baden-Fuller, C., Haefliger, S., 2013. Business models and technological innovation. Long. Range. Plann. 46 (6), 419–426.

Baden-Fuller, C., Morgan, M.S., 2010. Business models as models. Long. Range. Plann. 43 (2), 156–171.

Baden-Fuller, C., Winter, S.G., 2005. Replicating organizational knowledge: principles or templates? <http://dx.doi.org/10.2139/ssrn.1118013> (accessed 17.08.15).

Balci, O., 1998. Verification, validation, and accreditation. Proceedings of the 30th Winter Simulation Conference, Blacksburgh, VA, pp. 41–44.

Benlian, A., Hess, T., 2011. Opportunities and risks of software-as-a-service: findings from a survey of it executives. Decis. Support Syst. 52 (1), 232–246.

Benlian, A., Hess, T., Buxmann, P., 2009. Drivers of saas-adoption: an empirical study of different application types. Bus. Inf. Syst. Eng. 1 (5), 357–369.

Bharadwaj, A., El Sawy, O.A., Pavlou, P.A., Venkatraman, N., 2013. Digital business strategy: toward a next generation of insights. MIS Q. 37 (2), 471–482.

Boutellier, R., Eurich, M., Hurschler, P., 2010. An integrated business model innovation approach: it is not all about product and process innovation. Int. J. E-Entrepreneurship Innov. 1 (3), 1–13.

Brambilla, M., Fraternali, P., Vaca, C., 2012. BPMN and design patterns for engineering social bpm solutions. In: Daniel, F., Barkaoui, K., Dustdar, S. (Eds.), Business Process Management Workshops. Springer, Berlin, pp. 219–230.

Bughin, J., Chui, M., Manyika, J., 2010. Clouds, big data, and smart assets: ten tech-enabled business trends to watch. McKinsey Q. 4 (1), 26–43.

Burkard, C., Widjaja, T., Buxmann, P., 2012. Software ecosystems. Bus. Inf. Syst. Eng. 4 (1), 41–44.

Burkhart, T., Krumeich, J., Werth, D., Loos, P., 2011. Analyzing the business model concept – a comprehensive classification of literature. Proceedings of the 32nd International Conference on Information Systems, Shanghai, China, pp. 1–19.

Burkhart, T., Wolter, S., Schief, M., Krumeich, J., Di Valentin, C., Werth, D., et al., 2012. A comprehensive approach towards the structural description of business models. Proceedings of the International Conference on Management of Emergent Digital EcoSystems, Addis Ababa, Ethiopia, pp. 88–102.

Buschmann, F., Henney, K., Schmidt, D.C., 2007. Pattern-Oriented Software Architecture: On Patterns and Pattern Languages. John Wiley & Sons, Ltd, Chichester, UK.

Casadesus-Masanell, R., Ricart, J., 2010. From strategy to business models and onto tactics. Long. Range. Plann. 43 (2–3), 195–215.

Casadesus-Masanell, R., Ricart, J., 2011. How to design a winning business model. Harv. Bus. Rev. 89 (1–2), 100–107.

Casadesus-Masanell, R., Zhu, F., 2013. Business model innovation and competitive imitation: the case of sponsor-based business models. Strateg. Manage. J. 34 (4), 464–482.

Chesbrough, H., 2010. Business model innovation: opportunities and barriers. Long. Range. Plann. 43 (2–3), 354–363.

Chesbrough, H., Rosenbloom, R.S., 2002. The role of the business model in capturing value from innovation: evidence from xerox corporation's technology spin-off companies. Ind. Corp. Change 11 (3), 529–555.

Coyne, R., 1988. Logic Models of Design. Pitman, London.

Davidson, B., White, B.J., Taylor, M., 2012. The rise of it for entrepreneurs and the increasing entrepreneurial focus for it professionals. Issues Inf. Syst. 13 (2), 104–111.

Doganova, L., Eyquem-Renault, M., 2009. What do business models do?: innovation devices in technology entrepreneurship. Res. Policy 38 (10), 1559–1570.

Drucker, P.F., 1984. The discipline of innovation. Harv. Bus. Rev. 63 (3), 67–72.

Drucker, P.F., 1994. The theory of business. Harv. Bus. Rev. 72 (9), 95–104.

Drucker, P.F., 1999. Beyond the information revolution. Atl. Mon. 284 (4), 47–59.

Dunbrack, L.A., 2011. Connected Health Consumer Survey. IDC Health Insights, Framingham, MA.

Eisenmann, T., Parker, G., Van Alstyne, M.W., 2006. Strategies for two-sided markets. Harv. Bus. Rev. 84 (10), 92.

Enkel, E., Mezger, F., 2013. Imitation processes and their application for business model innovation: an explorative study. Int. J. Innov. Manage. 17 (1), 1–34.

Falcon, A., 2011. Aristotle on causality. Stanford Encyclopedia of Philosophy. <http://plato.stanford.edu/entries/aristotle-causality> (accessed 17.08.15).

Frankenberger, K., Weiblen, T., Gassmann, O., 2013. Network configuration, customer centricity, and performance of open business models: a solution provider perspective. Ind. Mark. Manage. 5 (42), 671–682.

Freeman, C., Louçã, F., 2001. As Time Goes By: From the Industrial Revolutions to the Information Revolution. Oxford University Press, Oxford, UK.

Gamma, E., Helm, R., Johnson, R., Vlissides, J., 1995. Design Patterns: Elements of Reusable Object-Oriented Software. Addison-Wesley Professional, Reading, MA.

Girotra, K., Netessine, S., 2011. How to build risk into your business model. Harv. Bus. Rev. 89 (5), 100–105.

Giudice, M.D., Straub, D. (Eds.), 2011. Editor's comments: it and entrepreneurism: an on-again, off-again love affair or a marriage? MIS Q. 35 (4), iii–viii.

Gordijn, J., Akkermans, H., 2001. Designing and evaluating e-business models. IEEE Intell. Syst. 16 (4), 11–17.

Gordijn, J., Akkermans, J., 2003. Value-based requirements engineering: exploring innovative e-commerce ideas. Requirements Eng. 8 (2), 114–134.

Gregor, S., Jones, D., 2007. The anatomy of a design theory. J. Assoc. Inf. Syst. 8 (5), 312–335.

Gregor, S., Martin, M., Fernandez, W., Stern, S., Vitale, M., 2006. The transformational dimension in the realization of business value from information technology. J. Strateg. Inf. Syst. 15 (3), 249–270.

Hackney, R., Burn, J., Salazar, A., 2004. Strategies for value creation in electronic markets: towards a framework for managing evolutionary change. J. Strateg. Inf. Syst. 13 (2), 91–103.

Hedman, J., Kalling, T., 2003. The business model concept: theoretical underpinnings and empirical illustrations. Eur. J. Inf. Syst. 12 (1), 49–59.

Hunter, L., 2006. Low cost airlines: business model and employment relations. Eur. Manage. J. 24 (5), 315–321.

Iacob, M.E., Meertens, L.O., Jonkers, H., Quartel, D.A.C., Nieuwenhuis, L.J.M., Sinderen, M.J., 2012. From enterprise architecture to business models and back. Softw. Syst. Model. 13 (3), 1059–1083.

Järvinen, P., 2007. On reviewing of results in design research. Proceedings of the 15th European Conference on Information Systems, St. Gallen, Switzerland, pp. 1388–1397.

Johnson, M.W., Christensen, C.M., Kagermann, H., 2008. Reinventing your business model. Harv. Bus. Rev. 86 (12), 57–68.

Kagermann, H., Österle, H., Jordan, J.M., 2010. IT-Driven Business Models: Global Case Studies in Transformation. John Wiley & Sons, Inc, Hoboken, NJ.

Katz, M.L., Shapiro, C., 1985. Network externalities, competition, and compatibility. Am. Econ. Rev. 75 (3), 424–440.

Keegan, P., 2009. Zipcar-the best new idea in business. <http://money.cnn.com/2009/08/26/news/companies/zipcar_car_rentals> (accessed 17.08.15).

Kiani, B., Shirouyehzad, H., Bafti, F.K., Fouladgar, H., 2009. System dynamics approach to analysing the cost factors effects on cost of quality. Int. J. Qual. Reliab. Manage. 26 (7), 685–698.

Kinder, T., 2002. Emerging e-commerce business models: an analysis of case studies from west lothian, scotland. Eur. J. Innov. Manage. 5 (3), 130–151.

Klang, D., Wallnöfer, M., Hacklin, F., 2014. The business model paradox: a systematic review and exploration of antecedents. Int. J. Manage. Rev. 16 (4), 454–478.

Kundisch, D., John, T., Honnacker, J., Meier, C., 2012. Approaches for business model representation: an overview. Proceedings of the Multikonferenz Wirtschaftsinformatik Braunschweig, Germany, pp. 1839–1850.

Lieberman, M.B., Montgomery, D.B., 1988. First-mover advantages. Strateg. Manage. J. 9 (S1), 41–58.

Lindgardt, Z., Reeves, M., Stalk, G., Deimler, M.S., 2009. Business Model Innovation: When the Game Gets Tough, Change the Game. The Boston Consulting Group Report, Boston, MA.

Lukka, K., 2003. The constructive research approach. In: Ojala, L., Hilmola, O.-P. (Eds.), Case Study Research in Logistics. Publications of the Turku School of Economics and Business Administration, Turku, pp. 83–101.

Mackenzie, I., Cohn, D., Gann, D., 2014. The new patterns of innovation. Harv. Bus. Rev. 92 (1-2), 86–95.

March, S.T., Smith, G.G., 1995. Design and natural science research on information technology. Decis. Support Syst. 15 (4), 251–266.

McGrath, R.G., 2010. Business models: a discovery driven approach. Long. Range. Plann. 43 (2-3), 247–261.

Mettler, T., 2013. Towards a unified business model vocabulary: a proposition of key constructs. J. Theo. Appl. Electron. Commer. Res. 9 (1), 19–27.

Mettler, T., Eurich, M., 2012. A "design-pattern"-based approach for analyzing e-health business models. Health Policy Technol. 1 (2), 77–85.

Morris, M., Schindehutte, M., Allen, J., 2005. The entrepreneur's business model: toward a unified perspective. J. Bus. Res. 58 (6), 726–735.

Neely, A., 2008. Exploring the financial consequences of the servitization of manufacturing. Oper. Manage. Res. 1 (2), 103–118.

Nunamaker, J.F., Chen, M., Purdin, T.D.M., 1990. Systems development in information systems research. J. Manage. Inf. Syst. 7 (3), 89–106.

Osterwalder, A., 2004. The business model ontology – a proposition in a design science approach, University of Lausanne.

Osterwalder, A., Pigneur, Y., 2002. An e-business model ontology for modeling e-business. Proceedings of the 15th Bled Electronic Commerce Conference, Bled, Slovenia, pp. 17–19.

Osterwalder, A., Pigneur, Y., 2010. Business Model Generation: A Handbook for Visionaries, Game Changers, and Challengers. John Wiley & Sons, Inc, Hoboken, NJ.

Osterwalder, A., Pigneur, Y., 2013. Designing business models and similar strategic objects: the contribution of IS. J. Assoc. Inf. Syst. 14 (5), 237–244.

Parker, G.G., Van Alstyne, M.W., 2005. Two-sided network effects: a theory of information product design. Manage. Sci. 51 (10), 1494–1504.

Pateli, A.G., Giaglis, G.M., 2004. A research framework for analysing ebusiness models. Eur. J. Inf. Syst. 13 (4), 302–314.

Peinel, G., Jarke, M., Rose, T., 2010. Business models for egovernment services. Electron. Gov. Int. J. 7 (4), 380–401.

Penker, M., Eriksson, H.-E., 2000. Business Modeling with UML: Business Patterns at Work. Wiley, New York.

Picker, R.C., 2010. The razors-and-blades myth(s). <http://www.law.uchicago.edu/files/file/532-rcp-razors.pdf> (accessed 17.08.15).

Pohle, G., Chapman, M., 2006. IBM's global CEO report 2006: business model innovation matters. Strategy Leadersh. 34 (5), 34–40.

Pree, W., 1994. Design Patterns for Object-Oriented Software Development. Addison-Wesley, Reading, MA.

Pynnonen, M., Hallikas, J., Savolainen, P., 2008. Mapping business: value stream-based analysis of business models and resources in information and communications technology service business. Int. J. Bus. Syst. Res. 2 (3), 305–323.

Rai, A., Tang, X., 2014. Research commentary-information technology-enabled business models: a conceptual framework and a coevolution perspective for future research. Inf. Syst. Res. 25 (1), 1–14.

Rappa, M., 2004. The utility business model and the future of computing services. IBM Syst. J. 43 (1), 32–42.

Sabatier, V., Mangematin, V., Rousselle, T., 2010. From recipe to dinner: business model portfolios in the european biopharmaceutical industry. Long. Range. Plann. 43 (2), 431–447.

Sako, M., 2012. Business models for strategy and innovation. Commun. ACM. 55 (7), 22–24.

Samavi, R., Yu, E., Topaloglou, T., 2009. Strategic reasoning about business models: a conceptual modeling approach. Inf. Syst. e-Bus. Manage. 7 (2), 171–198.

Schief, M., 2014. Business Models in the Software Industry. Gabler, Wiesbaden.

Seelos, C., Mair, J., 2007. Profitable business models and market creation in the context of deep poverty: a strategic view. Acad. Manage. Perspect. 21 (4), 49–63.

Shapiro, C., Varian, H., 1999. Information Rules: A Strategic Guide to the Network Economy. Harvard Business Press, Boston, MA.

Slywotzky, A.J., 1996. Value Migration: How to Think Several Moves Ahead of the Competition. Harvard Business Press, Boston, MA.

Sosna, M., Trevinyo-Rodríguez, R.N., Velamuri, S.R., 2010. Business model innovation through trial-and-error learning: the naturhouse case. Long. Range. Plann. 43 (2-3), 383–407.

Takeda, H., Veerkamp, P., Tomiyama, T., Yoshikawa, H., 1990. Modeling design processes. AI Mag. 11 (4), 37–48.

Teece, D.J., 2010. Business models, business strategy and innovation. Long. Range. Plann. 43 (2-3), 172–194.

Timmers, P., 1998. Business models for electronic markets. Electron. Mark. 8 (2), 3–8.

Veit, D., Clemons, E., Benlian, A., Buxmann, P., Hess, T., Kundisch, D., et al., 2014. Business models: an information systems research agenda. Bus. Inf. Syst. Eng. 56 (1), 45–53.

vom Brocke, J., 2007. Design principles for reference modelling. Reusing information models by means of aggregation, specialisation, instantiation, and analogy. In: Fettke, P., Loos, P. (Eds.), Reference Modelling for Business Systems Analysis. Idea Group Publishing, Hershey, PA, pp. 47–75.

von Krogh, G., Spaeth, S., 2007. The open source software phenomenon: characteristics that promote research. J. Strateg. Inf. Syst. 16 (3), 236–253.

Wooldridge, D., Schneider, M., 2010. The Business of iPhone App Development. Springer, New York.

Yip, G.S., 2004. Using strategy to change your business model. Bus. Strategy Rev. 15 (2), 17–24.

Yu, E., Giorgini, P., Maiden, N., Mylopoulos, J. (Eds.), 2011. Social Modeling for Requirements Engineering. MIT Press, Cambridge, MA.

Yunus, M., Moingeon, B., Lehmann-Ortega, L., 2010. Building social business models: lessons from the grameen experience. Long. Range. Plann. 43 (2), 308–325.

Zott, C., Amit, R., 2010. Business model design: an activity system perspective. Long. Range. Plann. 43 (2–3), 216–226.

Zott, C., Amit, R., Massa, L., 2011. The business model: recent developments and future research. J. Manage. 37 (4), 1019–1042.

CHAPTER 10

Business Models in Two-Sided Markets (Analysis of Potential Payments and Reimbursement Models That Can Be Used)

V. Vimarlund[1,2] and T. Mettler[3]
[1]Linköping University, Linköping, Sweden
[2]Jönköping University, Jönköping, Sweden
[3]University of Lausanne, Lausanne, Switzerland

INTRODUCTION

As understood from prior research and discussed in previous chapters of this book, two-sided markets are driven by network economics and complementary product pricing and, as such, focusing on indirect networks effects and the importance of balancing the interests and needs from both sides of the market (Rochet and Tirole, 2006). Accordingly, the utility on one side of the market increases with the number (and/or quality) of participants on the other side. There has been a plethora of research describing the functioning of two-sided markets, particularly from the lens of the private sector (Armstrong, 2006; Caillaud and Jullien, 2003; Evans, 2003; Rochet and Tirole, 2003). A focal point of attention in previous work has been on strategies for how intermediaries can generate profits while still creating some kind of value for their platform members (Eisenmann et al., 2006; Gawer and Cusumano, 2008; Hidding et al., 2011), how to overcome the "competitive bottleneck" of catching service consumers by charging them less than cost in order to attract paying service providers (Armstrong and Wright, 2007), as well as more pragmatically, how to set up such a platform to harness the benefits of either open or closed platform designs (Tåg, 2008). However, previous work frequently neglects the particularities of intermediary platforms, which are operating in nontransaction-oriented or highly regulated markets, such as e-health. Particularly for e-health, business models, and their respective payment and reimbursement schemes often need to follow an alternative logic in order to stimulate actors to join the market and to move from a first stage (in which the market is tested) to a more sustainable situation where intermediaries do not sell themselves out and a balance of interests is established.

In recent years, we have seen many public and private organizations, such as local or regional authorities, clinics or insurance or pharmaceutical companies, developing certain types of intermediary platforms. Due to policies and regulations of the health system, they are operating in, many of them have become a "gate keeper" (Baldwin and Clark, 2006), deciding who is given access to the platform, and what is been offered to whom. Far less have taken up the role of "invisible engine" that allows for a more profound digital transformation and renewal of the health system by letting SME and entrepreneurs combine their core competencies with complementary products and service offerings, influencing in this manner organizational strategies as well as consumer behavior and suppliers preferences.

We believe that while there is a certain advantage in being first mover, such as having an established customer base, switching costs, and network effects, the e-health market also needs successful followers that provide complementary products in order to make intermediary platforms sustainable. Platforms owners and decision makers should therefore pay close attention to how they attract and manage their complementors by supporting openness of the platform, reducing the tension between innovation and efficiency, and finally by identifying innovative business, payment, and reimbursement models.

In this chapter we therefore center our attention on different types of business models and corresponding revenue and reimbursement schemes for intermediary platforms of e-health services. The identification of business and reimbursement models in two-sided e-health markets has normally followed the models and concepts used in the private sector. Intermediary platforms have been assumed to have the power to create niche markets. Which kind of business model will be applied by intermediary platforms, and will much depend on the role the platform and the level of control the platform owner has over demand membership, subscription, or any other kind of fees that one of the two groups needs to pay to be able to be a part of the market? In what follows, we first explain what we understand under the term "business model" and then describe the components of different business models for intermediary platforms for a two-sided e-health market, applying theory to practice and using as example a health account in development.

BUSINESS MODELS IN TWO-SIDED E-HEALTH MARKETS

Business models have been a focus for both practitioners and academics since the expansion of the Internet during the 1990s (Zott et al., 2011). Since then, many deliberations about business models have been made, ranging in ideas and concepts of how to design, develop, evaluate, and deploy such models in a digital setting. Today, one area of major interest is the role of service innovation as enabler for novel business models (Barrett et al., 2015; Lusch and Nambisan, 2015), as it has proven to radically transform traditional industries such as travel, news media, entertainment, or (as emphasized in this book) healthcare.

In Europe, several countries have developed strategies to foster the use and delivery of e-health[1] services through platforms and portals (European Commission, 2012). In these strategies, business models are perceived as a centerpiece and part of policy work that aims to support the design and deployment of e-health services (Moen et al., 2012). Prior research on e-health business models has mostly focused on healthcare settings where reimbursements came from patients' or employers' insurance companies or from social security (Kijl et al., 2010; Van Ooteghem et al., 2012; Visser et al., 2010). A more recent review focused on multipayer or market-based financial systems (Acheampong and Vimarlund, 2015). There are, however, few studies that discuss business models, their pros, and cons, and the components that are of relevance for intermediary platforms in a two-sided e-health market.

Before we delve into particular considerations on how to develop intermediary platforms for two-sided e-health markets, let us explain our understanding of the term "business model" first. Previous research has shown that the concept of business models is defined in different ways to suit the purpose of each study (Al-Debei and Avison, 2010; Osterwalder and Pigneur, 2013; Zott et al., 2011). The lack of consensus has been connected to the novelty of the concept (Osterwalder, 2004), its multidisciplinary background (Pateli and Giaglis, 2004), and the innovative areas where the concept is investigated and applied (Al-Debei and Avison, 2008). In the body of extant literature, we found various definitions of the concept (Rappa, 2003; Shafer et al., 2005; Timmers, 1998), descriptions of what constitutes a business model (Dubosson-Torbay et al., 2002; Gordijn and Akkermans, 2003; Hedman and Kalling, 2003; Osterwalder and Pigneur, 2010), its relation to strategy and innovation (Chesbrough, 2010; Teece, 2010), as well as various classifications and/or frameworks for differentiating business models (Al-Debei and Avison, 2010; Pateli and Giaglis, 2004; Zott et al., 2011). In addition to the lack of consensus concerning the definition of the business model concept, there seems to also be a lack of consensus when describing the number of components that a business model should include (Mettler, 2014).

To our view, a business model—per definition—refers to key organizational aspects, such as business activities, company resources and capabilities, financial, social, and physical structures and the like, as well as comprising not only the market side but also units operating, from product or service design through manufacture, distribution channels, and to the customers. Business models for nonprofit and/or institutional organizations are often associated with reimbursement or revenue models, and are used when a business opportunity appears, when innovations, such as e-health services, need to be integrated and financed, or when there is a strong need for developing pricing alternatives for information or digital services. An issue when developing business models for nonprofit or public organizations is how to guarantee the sustainability of the model in markets in which customers are normally segmented by their condition, expertise, product type or usage, and by its geographical location. Further, in markets like the ones we focus on in this book, there are even other issues of relevancy, namely that there is insufficient information about the customers and their preferences and about how business models will influence the supply chain and interaction between different stakeholders.

Revenue models are a subset of the larger business models. They are not mutually exclusive, and often work interchangeably in all business models depending on the specific context and requirements of the business. An important difference between business and revenue models is that while revenue models has often a selling perspective and implicit includes prices for services or products, business models focus on how organizations create, deliver, and share values between stakeholders in a sustainable manner (Eurich et al., 2011).

The application of business models and revenue models to the area of e-health has been heavily influenced by the prior-mentioned work on business models in e-commerce and other industries. Studies about business models in e-health are diverse and cover different aspects as to the design, evaluation, and deployment to healthcare practice (Kimble, 2015; Valeri et al., 2010). For instance, different frameworks for analyzing and conceptualizing business models for distinct e-health services exist (Chen et al., 2013; Lin et al., 2010; Lin et al., 2011; van Limburg et al., 2011; Visser et al., 2010) or different guidelines, best practices, and considerations of how to develop sustainable revenue models and how to overcome market barriers (Mettler and Eurich, 2012; Parente, 2000; Spil and Kijl, 2009; Sprenger and Mettler, 2016). However, there has not yet been much evidence for how business models can be applied, and their relevance to two-sided e-health markets so far and the important role for intermediaries, from newcomers to established actors, within such a market, is to what we now turn.

THE ROLE OF THE INTERMEDIARY PLATFORM

Many intermediary platforms in two-sided e-health markets have mimicked the example of Apple and Google, forming coalitions and trying to create niche markets. The underlying idea of these platforms has therefore been driven by the urge to attract service providers that produce complementary e-health services for specific groups of service consumers, such as general practitioners or patients, and to develop an operational structure that allows them to obtain some kind of cross-side-effect. Intermediary platform owners thus frequently follow the motto: "The more developers join a platform, the higher is our users' benefit." To attract service providers platform owners not only are required to offer a proper infrastructure (e.g., development kits, application programming interfaces, data storage) and to define rules that clearly describe how business is done, more importantly they also need to provide the necessary structure and information about the market (e.g., amount of specific services per category, usage time or downloads by service consumers) such that they can act as "connectors" between sellers and buyers, respectively, serve as a trusted third party by controlling and/or certifying the quality of the available offerings and service providers.

There are, however, some major issues that intermediary platform owners are confronted with. One of them is "single-homing" behavior driven by requirements and/or imposed by exclusive contracts, or crowding-out effects of increasing competition (the more developers participate on a platform, the lower the attention to a single developer becomes). The mentioned issues significantly influence the sustainability of the two-sided market and consequently the business and reimbursement models applied. Understanding the impact of these issues is of key relevance for addressing the challenges pertaining to the role of an intermediary within two-sided e-health markets. Accordingly, owners of intermediary platforms in two-sided markets need to reflect on strategic alternatives, frequently not present in single-sided markets. From a strategic perspective, it is necessary to elucidate: (i) the "added-value" or in other words if the platform performs one or several essential functions or solves one or several essential problems for several actors, (ii) the "transition" respectively if there is a need for progression between reimbursement and payment models during the evolution from an emerging to a well-established market, and (iii) the "viability-fit-ratio" respectively the risks and challenges that every component of the business model has in terms of technological, economic, organizational, and societal considerations (Mettler, 2016). In the next section, we thus delineate a concrete case to exemplify how to integrate this strategic perspective into the business model design for two-sided e-health markets.

APPLYING THEORY TO PRACTICE: AN EXAMPLE OF BUSINESS AND REVENUE MODELS FOR AN E-HEALTH INTERMEDIARY PLATFORM

The case we want to highlight in this book chapter is the national health account HealthForMe (Fig 10.1), which is an initiative from the Swedish Ministry of Health and Social Affairs that seeks to offer all citizens an intermediary platform (HealthForMe) where they interact with services providers and use a wide variety of e-health services (eHälsomyndigheten, 2016). In this sense, the intermediary platform is meant to be a "meeting place" for both developers of innovative e-health services and patients, which seek to get an overview, manage, and further exploit their health information with digital services. The Swedish e-health agency is responsible for the public procurement and implementation of the account, with a budget of 26 million EUR between 2014 and 2017. Each citizen will own the information that is gathered in the health account and can, whenever he or she wishes, decide to delete or share the information with family members, healthcare professionals, or any other third party which is on the platform.

The account will support exchanges between "buyers" and "sellers." Public and private organizations, SMEs, and entrepreneurs will have the possibility to offer a series of e-health services and apps connected to the accounts of users. The intermediary platform is aimed to provide app developers access to a sophisticated development environment, which should facilitate the fast deployment of applications and reduce duplicate costs.

Buyers of services will be able to search and find participants on the other side and have the opportunity to consummate matches. Having a large number of participants on both sides will increase the probability that participants will find a match. Third-party developers that want to connect and offer services to citizens will need to apply to the Swedish e-health Agency to be certified according to ethical and technical requirements.

It is expected that the intermediary platform will have a considerable societal impact as it will support different groups of customers in accessing and using e-health services and allow a member of one group (developers) to benefit from having his services and apps coordinated with one or more members of another group. Finally, a

Figure 10.1 HealthForMe Website *eHälsomyndigheten. 2016. Retrieved (12.07.16), from https://www. halsaformig.se/.*

major efficiency boost is gained from coordination (rather than bilateral relationships) between the members of the groups.[2] Citizens will consequently be able to sample and share information from electronic health records (EHRs), vaccinations, medications lists, and the like. They will also have the possibility to download and use apps and services delivered by third-party organizations, which focus on broader areas of health and wellbeing, such as digital prevention or fitness offerings. Overall, the intermediary platform is expected to stimulate citizens' engagement in their own health, strengthen patient involvement in healthcare, and create the prerequisites for new e-health services and alternative forms of cure and care activities.

The provision of a national health account to all citizens, however, brings challenges that have to be managed in order to fulfill an effective delivery of the services offered on the platform. Some of the challenges depend on the expectations of entrepreneurs, others on the expectations of users or citizens, and others on the evidence-based experiences of care providers as well as on issues related to the long-term ownership of the platform.

In order that HealthForMe can be successful, all components of its business model need to fit together in a cooperative and supportive fashion. As we described previously, a business model contains various aspects such as the description of the main business activities, company resources and capabilities, financial, social, and physical structures. Besides an in-depth delineation of the mentioned components, a successful business model should also contain an analysis of consumer purchase intentions, concretizing the extent to which customers are likely to purchase the offered e-health services and how much they are willing to pay for it. In the case of HealthForMe the potential stakeholders as well as the customers and the owner of the intermediary platform have been previously identified as citizens, researchers, healthcare providers (as service consumers), public and private organizations (as services providers), and the Swedish e-health Agency (as platform owner).

It is important to note that the business model should be specific as to the goal it wants to achieve; for instance, HealthForMe wants to be an matchmaker in the Swedish health system, fostering interaction, access, and exchange between consumers and providers of e-health services that share common interests. The objective is to capture a major part of the market, and have revenues that cover the costs for the design, development, implementation, and governance of the platform in ten years from now. Whether these goals and objectives are or can be realistic or not for HealthForMe, and whether the organization that owns the platform is prepared to achieve these goals, needs to be developed and evaluated iteratively.

For the purpose of exemplifying at least a particular instant of this iterative process, we will now describe the components of the business model for HealthForMe. We limit the description in Table 10.1 to the identification of the different components the model should have (a projection of the to-be state; as opposed to the current as-is

Table 10.1 Example of how to describe components and challenges of the projected business model for *HealthForMe*

Component	Description
Actors	• Swedish e-health Agency (platform owner) • Researchers, healthcare providers, or other public organizations (service provider and consumer) • Entrepreneurs, SMEs, and other private organizations (mainly service provider) • Citizens (mainly service consumer) Challenge: How "powerful" will the owner of the platform be, and which other partners will be allowed to join the "ownership group" (e.g., local and regional governmental authorities)? Which other groups of actors are allowed to enter into the market (e.g., patient organizations, private insurance offices, private healthcare providers)? How to handle eventual regulations, policies, or laws that influence the delivery of the service in cases of organizations and individuals outside the healthcare system that will use the services? Who is ultimately responsible for failures or violation to actual rules, policies, or law?
Activities	• Sample, delivery, and transfer of secure data • Ensure that data delivered achieves rich quality demands • Delivery of information between and within specialized organizations outside healthcare contexts Challenge: What happens if collaboration between two or more service providers is developed with the aim to provide customers more than a simple solution (e.g., a platform within the existing platform)? How to ensure that information that is changing over the time will be actualized in the platform? Where does the responsibility of the platform for renewal of information begin and end?
Resources	• Data from doctor-patient encounters, bio bank databases, professional health services like vaccination apps • Data sampled by individuals (e.g., quantified-self) or non-specialized organizations Challenge: Has the intermediary platform owner the necessary resources to (financially) maintain the platform in the future? Does the intermediary platform owner have the necessary personnel resources to administrate and innovate the platform? Who will account for the administrative costs of service maintenance?
Customer relationship	• Technical platform • Service desk, telephone hotline, other support services • Branding and marketing of platform Challenge: Should there be one single point of contact or several coexisting ones? Who will take the responsibility for liaising with the customer? Do the providers offer a secure channel for distribution of information?

(Continued)

Table 10.1 Example of how to describe components and challenges of the projected business model for *HealthForMe* Continued

Component	Description
Cost structure and revenue stream	• Fix (e.g., salaries) and variable costs for operating and further developing the platform • Direct (e.g., admission fee, download fees) and indirect revenues (e.g., advertisements, affiliate services) from operating the platform Challenge: Should marketing costs, governance costs, and administrative costs for the platform be included and charged to both or only one side of the market? Should the platform apply the fast or/and non-variable costs principle over the years and charge platform users based on it? How to formalize accountability and how to transfer reimbursement between the different stakeholders, if needed? How to build compensation mechanisms in case of potential brand manage if one of the partners provides less than expected quality in the service?
...	...

state), and to the description of issues that need to be clarified to ensure the sustainability of the business model in the future.

In an ordinary one-sided market under the usual assumptions, buyers' willingness to pay for incremental units of output provides a measure of the social value of the intermediary platform. Thus the social optimum in such markets occurs at the output level at which price is equal to marginal cost, since at lower levels of output buyers are willing to pay more than marginal cost for incremental output, while at higher levels of output they are willing to pay less than marginal cost. What is different in two-sided e-health markets is the fact that the platform used to distribute information and services (e.g., vaccines, personal health records) is not particularly vulnerable to competition as compared to other consumer areas. Furthermore, actors in the e-health market are often vertically integrated and provide information services that are not produced by firms or companies but by healthcare organizations. In this context, there is often an absence of transactions, which leads to a situation in which there is not only a buyer and a seller of information, but also a group of individuals that voluntary give their data for free (i.e., a patient that accepts that data and information available in his/her journal are re-used) and groups of individuals and/or organizations that will use individuals' data in their services, from which they will be paid for by other individuals.

In general, a business model for an intermediary platform, such as HealthForMe, needs to have the flexibility to be adapted to different situations and the possibility to use different kinds of revenue streams depending on the character of transaction or information exchange as well as the actor to whom the service is rendered. Eurich et al. (2011) describe different direct and indirect mechanisms of recurring and non-recurring revenues. In Table 10.2 we provide some illustrative examples of well-known

Table 10.2 Description of different revenue models of potentially interest for *HealthForMe*

Type of model	Description
Brokerage model	• An intermediary platform that uses the brokerage model acts as a market-maker, i.e., they bring buyers (e.g., patients) and sellers (e.g., e-health apps developers) together, and facilitate transactions. Brokers play a frequent role in business-to-business (B2B), business-to-consumer (B2C), or consumer-to-consumer (C2C) markets. Usually a broker charges a fee or commission for each transaction it enables. • Brokerage models demand the existence of a (1) marketplace exchange that offers a full range of services covering the transaction process, from market assessment to negotiation and fulfillment, (2) buy/sell fulfillment that takes customer orders to buy or sell a product or service, including terms like price and delivery, (3) transaction procedures to provide a payment mechanism for buyers and sellers to settle a transaction, and (4) a search agent, for instance, a software agent used to search out the price and availability for a service specified by the buyer.
The producer model	• An intermediary platform using the producer model allows a service provider (in our case for example a healthcare professional) to reach buyers directly and thereby compress the distribution channel. The producer model can be based on efficiency, improved customer service, and a better understanding of customer preferences. • The producer model presupposes that (1) the sale of a product or service implies that the right of ownership is transferred to the buyer, (2) the service cannot be returned to the seller upon expiration or default. It can, however, be re-used, or transferred by the buyer in several other contexts, geographical areas, once the property rights of the service has been transferred to the buyer.
Community model	• The viability of the community model is based on user loyalty. Users have a high investment in both time and emotion. Revenue can be based on the sale of products and services or voluntary contributions, or revenue may be tied to contextual advertising and subscriptions for premium services. Digital services are inherently suited to community business models. • The most common examples of community models are social networking services. They provide individuals with the ability to connect to other individuals along a defined common interest (professional or leisure). They are normally developed as open source, are accessible by a global community of contributors, and are usually not-for-profit. The community is often supported by voluntary donations.

revenue models, which can be applied in settings like our HealthForMe case. It is important, however, to note that the choice of the model depends on various considerations, such as (1) the importance of creating opportunities to attract new customer segments, (2) the level of dependence between the platform owner and the corresponding service providers, (3) the geographic location or statutory requirements and expectations with regard the service delivery to consumers.

CONCLUSION

There are a variety of components that influence whether or not a business model is profitable, including location, leadership, market demand, competition, and so on. This is no different for business models of an intermediary platform in two-sided e-health markets. To our knowledge, there is scarce literature analyzing the business models of such platforms, how much money potential customers are willing to pay for e-health services, or even the fact of how many customers are actually willing to use such a platform for managing and exchanging their personal health information and consuming e-health services from both health professionals and third party organizations. The absence of information is a mayor constraint for the identification and estimation of sources of revenue, total costs or price of the e-health services to be developed.

In this book chapter, we showed the example of HealthForMe and described many of today's open questions and challenges to be addressed by an intermediary operating in a two-sided e-health market. We have seen that business models for such intermediaries require a strategic perspective on challenges, which are frequently not present in single-sided markets. Our analysis points out that the key implication for intermediary platforms is to push further for consolidation while possible challengers have to look for their niche to be successful. For developers it is crucial to scan the market for disruptions and trends toward consolidation in order to efficiently allocate their resources. To support developers, we suggest intermediaries of two-sided e-health markets focus on the following:

- Identification of alternative revenue models and pricing strategies that encourage internal and external innovation and foster a fruitful relationship with complementors
- Research how price-sensitive potential users of e-health services are and how they react to actions of competing platforms
- Development and application of approaches and models for eliciting the amount of fees and costs for joining or consuming certain services of the platform
- Creation of strategic alliances with institutional, nonprofit, and for-profit organizations in order to extend the service offering and attractiveness of the platform
- Definition or adjustment of policies and regulations which prevent governance, intellectual property, and other types of issues

ENDNOTES

1. Following many prominent authors, "e-health" (or the delivery of health services and information through the Internet) does not only denote a technical development, but also a new way of working to improve healthcare locally, regionally, and worldwide by using information and communication technology (Eysenbach, 2001; Pagliari et al., 2005).
2. Information and transaction costs and free-riding problems make it difficult for members of distinct customer groups to internalize the externalities on their own and to make entry into bilateral transactions.

REFERENCES

Acheampong, F., Vimarlund, V., 2015. Business models for telemedicine services: a literature review. Health Syst. 4 (3), 189–203.

Al-Debei, M.M., Avison, D., 2008. Defining the business model in the new world of digital business. Proceedings of the Americas Conference on Information Systems (AMCIS), pp. 1–11.

Al-Debei, M.M., Avison, D., 2010. Developing a unified framework of the business model concept. Eur. J. Inf. Syst. 19 (3), 359–376.

Armstrong, M., 2006. Competition in two-sided markets. RAND J. Econ. 37 (3), 668–691.

Armstrong, M., Wright, J., 2007. Two-sided markets, competitive bottlenecks and exclusive contracts. Econ. Theory 32 (2), 353–380.

Baldwin, C.Y., Clark, K.B., 2006. Architectural innovation and dynamic competition: The smaller "footprint" strategy. Harvard Business School, Boston.

Barrett, M., Davidson, E., Prabhu, J., Vargo, S.L., 2015. Service innovation in the digital age: key contributions and future directions. MIS Q 39 (1), 135–154.

Caillaud, B., Jullien, B., 2003. Chicken & egg: competition among intermediation service providers. RAND J. Econ. 34 (2), 309–328.

Chen, S., Cheng, A., Mehta, K., 2013. A review of telemedicine business models. Telemed. e-Health 19 (4), 287–297.

Chesbrough, H., 2010. Business model innovation: opportunities and barriers. Long Range Plan. 43 (2–3), 354–363.

Dubosson-Torbay, M., Osterwalder, A., Pigneur, Y., 2002. E-business model design, classification, and measurements. Thunderbird Int. Bus. Rev. 44 (1), 5–23.

eHälsomyndigheten. 2016. Retrieved (12.07.16), from <https://www.halsaformig.se/>.

Eisenmann, T., Parker, G., Van Alstyne, M.W., 2006. Strategies for two-sided markets. Harvard Bus. Rev. 84 (10), 92.

Eurich, M., Giessmann, A., Mettler, T., Stanoevska-Slabeva, K., 2011. Revenue streams of cloud-based platforms: Current state and future directions, In: 7th Americas Conference on Information Systems. Detroit, Michigan: pp. 1–10.

European Commission. 2012. Ehealth action plan 2012–2020: innovative healthcare for the 21st century. Retrieved (12.07.16) from <https://ec.europa.eu/digital-single-market/en/news/ehealth-action-plan-2012-2020-innovative-healthcare-21st-century>.

Evans, D.S., 2003. Some empirical aspects of multi-sided platform industries. Rev. Netw. Econ. 2 (3), 191–209.

Eysenbach, G., 2001. What is e-health? J. Med. Internet Res. 3 (2), e20.

Gawer, A., Cusumano, M.A., 2008. How companies become platform leaders. MIT Sloan Manage. Rev. 49 (2), 27–35.

Gordijn, J., Akkermans, H.M., 2003. Value-based requirements engineering: exploring innovative e-commerce ideas. Requirements Eng. 8 (2), 114–134.

Hedman, J., Kalling, T., 2003. The business model concept: theoretical underpinnings and empirical illustrations. Eur. J. Inf. Syst. 12 (1), 49–59.

Hidding, G.J., Williams, J., Sviokla, J.J., 2011. How platform leaders win. J. Bus. Strategy 32 (2), 29–37.

Kijl, B., Nieuwenhuis, L., Huis in't Veld, R., Hermens, H., Vollenbroek-Hutten, M., 2010. Deployment of e-health services – a business model engineering strategy. J. Telemed. Telecare 16 (6), 344–353.

Kimble, C., 2015. Business models for e-health: evidence from ten case studies. Global Bus. Organ. Excell. 34 (4), 18–30.

Lin, S., Liu, J., Wei, J., Yin, W., Chen, H., Chiu, W., 2010. A business model analysis of telecardiology service. Telemed. J. E-health 16 (10), 1067–1073.

Lin, T., Chang, H., Huang, C., 2011. An analysis of telemedicine in Taiwan: a business model perspective. Int. J. Gerontol. 5 (4), 189–192.

Lusch, R.F., Nambisan, S., 2015. Service innovation: a service-dominant logic perspective. MIS Q.y 39 (1), 155–175.

Mettler, T., 2014. Towards a unified business model vocabulary: a proposition of key constructs. J. Theor. Appl. Electro. Commerce Res. 9 (1), 19–27.

Mettler, T., 2016. Anticipating mismatches of HIT investments: developing a viability-fit model for e-health services. Int. J. Med. Inform. 85 (1), 104–115.

Mettler, T., Eurich, M., 2012. A "design-pattern"-based approach for analyzing e-health business models. Health Policy Technol. 1 (2), 77–85.

Moen, A., Hackl, W.O., Hofdijk, J., Gemert-Pijnen, L.V., Ammenwerth, E., Nykänen, P., et al., 2012. Ehealth in Europe – status and challenges. Eur. J. Biomed. Inform. 8 (1), 2–7.

Osterwalder, A., 2004. The Business Model Ontology – A Proposition in a Design Science Approach. University of Lausanne.

Osterwalder, A., Pigneur, Y., 2010. Business Model Generation: A Handbook for Visionaries, Game Changers, and Challengers. Wiley, Hoboken, NJ.

Osterwalder, A., Pigneur, Y., 2013. Designing business models and similar strategic objects: the contribution of is. J. Assoc. Inf. Syst. 14 (5), 237–244.

Pagliari, C., Sloan, D., Gregor, P., Sullivan, F., Detmer, D., Kahan, J., et al., 2005. What is ehealth (4): a scoping exercise to map the field. J. Med. Internet Res. 7 (1), e9.

Parente, S.T., 2000. Beyond the hype: a taxonomy of e-health business models. Health Affairs 19 (6), 89–102.

Pateli, A.G., Giaglis, G.M., 2004. A research framework for analysing ebusiness models. Eur. J. Inf. Syst. 13 (4), 302–314.

Rappa, M. 2003. Managing the digital enterprise: business models on the web. Retrieved (24.07.2014) from <http://digitalenterprise.org/models/models.html>.

Rochet, J.-C., Tirole, J., 2003. Platform competition in two-sided markets. J. Eur. Econ. Assoc. 1 (4), 990–1029.

Rochet, J.-C., Tirole, J., 2006. Two-sided markets: a progress report. RAND J. Econ. 37 (3), 645–667.

Shafer, S., Smith, H., Linder, J., 2005. The power of business models. Bus. Horizons 48 (3), 199–207.

Spil, T., Kijl, B., 2009. E-health business models: from pilot project to successful deployment. IBIMA Bus. Rev. 1 (5), 55–66.

Sprenger, M., and Mettler, T. 2016. On the utility of e-health business model design patterns. In: 24th European Conference on Information Systems. Istanbul, Turkey, pp. 1–16.

Tåg, J., 2008. Essays on Platforms: Business Strategies, Regulation and Policy in Telecommunications, Media and Technology Industries. Hanken School of Economics, Helsinki.

Teece, D.J., 2010. Business models, business strategy and innovation. Long Range Plan. 43 (2–3), 172–194.

Timmers, P., 1998. Business models for electronic markets. Electron. Mark. 8 (2), 3–8.

Valeri, L., Giesen, D., Jansen, P., and Klokgieters, K. 2010. Business models for ehealth.

van Limburg, M., van Gemert-Pijnen, J., Nijland, N., Ossebaard, H., Hendrix, R., Seydel, E., 2011. Why business modeling is crucial in the development of ehealth technologies. J. Med. Internet Res. 13 (4), e124.

Van Ooteghem, J., Ackaert, A., Verbrugge, S., Colle, D., Pickavet, M., Demeester, P., 2012. Economic viability of ecare solutions. In: Szomszor, M., Kostkova, P. (Eds.), Electronic Healthcare: Third International Conference, ehealth 2010, casablanca, morocco, December 13–15, 2010, revised selected papers. Springer, Berlin, Heidelberg, pp. 159–166.

Visser, J.J.W., Bloo, J.K.C., Grobbe, F.A., Vollenbroek-Hutten, M.M.R., 2010. Video teleconsultation service: Who is needed to do what, to get it implemented in daily care? Telemed. e-Health 16 (4), 439–445. 2010/05/01.

Zott, C., Amit, R., Massa, L., 2011. The business model: recent developments and future research. J. Manag. 37 (4), 1019–1042.

PART VI

The Future of the Area

CHAPTER 11

The Future of Two-Sided E-Health Markets

V. Vimarlund[1,2]
[1]Linköping University, Linköping, Sweden
[2]Jönköping University, Jönköping, Sweden

"The biggest challenge of e-health is to move from research to deployment to market."

THE FUTURE NEEDS A MARKET PERSPECTIVE ON E-HEALTH

The two-sided e-health market is rapidly becoming fundamental for health and social care. Worldwide many different steps have been taken to increase the engagement of consumers with e-health, mainly focusing in the development of novel digital services that increase well-being or tackle some social challenges, such as the lack of qualified personnel, or dwindling resources. At the same time, some effort is also spent on establishing a market in which both sides—e-health consumers and providers—can interact and benefit from with each other (Connell and Young, 2007). In this sideline, there has been a great ambition to introduce "service innovation", "design thinking" and other tenors of the service-dominant logic (Vargo and Lusch, 2008) to open up for new collaborations between private and public actors. Intermediary platforms, provided by regional or national authorities (Aanestad and Jensen, 2011) or private actors such as insurance companies (Scott et al., 2006), become a key coordination infrastructure that allows information to flow within and between the two sides of the market, regulating nontransaction activities and making decisions that determine which group receives support and in which manner, and which kind of price structure will exist to stimulate the two sides to become an active actor of the market.

A characteristic of the two-sided e-health market is, however, that a relatively small number of firms have the capability and capacity to do so. Small innovative SMEs compete against big, sometimes extremely slow but powerful multinational conglomerates in this market. Typically, large firms have advantages over smaller firms, at least up to a point, because their larger size delivers more value—a bigger network—to consumers, and thus the possibility to benefit from network effects and the effects on the fixed costs for getting one or both sides onboard. There are, however, other factors that influence the opportunities to earn competitive profits (i.e., profits that exceed those necessary to attract capital to the industry after accounting for risk) that influence the market, limiting a large number of SME firms from entering and competing with their services against big multinationals. As the value of joining an intermediary platform depends on the installed base, or more generally the demand of customers of the other side, the benefit of using a platform depends on the requests for usage by the other side. Even in the case where no transaction-oriented two-sided markets are developed, either because of decisions from governmental authorities, policies, and regulations or social and ethical issues (Anderson, 2007), the absence of information on the demand of, for example, multihome alternatives, the existence of networks, price structures, and/or existence of information about preferences, influences the market and builds competitive constraints. Several of these issues have been discussed in the chapters included in this book, visualizing and explaining the pros and cons, but also disclosing experiences, which can be used to actively shape the future of e-health markets.

Notwithstanding the mentioned differences between SME and multinationals, the establishment of working and sustainable two-sided e-health markets faces important challenges for all types of organizations, professional groups, and consumers. As earlier chapters in this book have demonstrated, there is a need to develop innovative business models, discussing security and safety issues, defining guidelines and concepts to ensure the sustainability and implementation e-health services, and having a more in-depth understanding of how to use the information available on the market to further diffuse the developed e-health services and broaden the installed base.

The chapters in this book have additionally addressed many theoretical, practical, conceptual, and IT-related gaps, which are currently discussed when dealing with two-sided e-health markets. In our pursuit of better understanding how to establish a sustainable market, we have come to recognize that a multidisciplinary research approach is required, and that discussions need to broaden in terms of the scope of innovation, renewal of established structures, and critical examination of consumption behaviors. In addition, a dynamic perspective (e.g., over time, health systems, political reforms) is needed, as value expectations from two-sided markets develop and understandings and connotations of concepts (e.g., what is e-health?) evolve. Furthermore, it is imperative to discuss the importance of social innovation in the context of e-health, since this brings additional dynamics in customer demand and may radically influence the relevance of a faster time-to-market or yield to increased competition and establishment of cocreation in value networks.

In view of the challenges a two-sided e-health market is confronted with to achieve sustainability, this book contributes to a rethinking of the relevance of the market perspective and the obligation to not only spend time with medical or technical deliberations but also with studying the effects on the usage and diffusion of e-health services, as well as the importance of developing the adequate societal and political support; in particular with formulating policies and regulations that support the innovativeness of the market and renewal of the organizational and/or financial structures of the health system.

It is therefore important to move from a stage of theoretical discussion to practical action or as we would like to emphasize, *"to move from research to deployment to market."* It is about time to assess how intermediary platforms owned by public or private organizations will face their responsibility to ensure safety, efficacy, efficiency, and at the same time, to support technological and organizational innovation which enable concrete advances of two-sided markets. Contrariwise, the sustainability of the market depends on a number of factors which are not directly dependent on the intermediaries. Accordingly, it is important to also focus the attention on issues, such as trust or ability of data gathering, guidelines and rules of data privacy and security, and intertemporal consumer behavior.

Earlier chapters in this book have indicated a couple of important issues related to both market and nonmarket related future topics, which we will summarize below.

THE ROLE OF THE INTERMEDIARY PLATFORM AND ITS IMPLICATIONS IN INCREASING INDUSTRY, SME, AND ENTREPRENEURIAL INVOLVEMENT

One of the most important issues to stimulate the private sector to become a part of the market is the introduction of intermediary platforms for accelerating diffusion of innovation, for shortening cycle times and for actively promoting alternative ideas and thinking. A mayor difficulty for the e-health market today is that many innovations have no defined end-point. Innovations in e-health are often incremental and form part of a continuous process of numerous small-scale advances. A pragmatic approach to the implementation and diffusion of e-services in a two-sided context needs to be developed.

Further, an intermediary platform will depend on the manner in which it is incorporated and used by practitioners, experts, suppliers, and customers. An important challenge will be to link the information sampled by the different actors interacting in the platform for clinical decisions, either directly (through feedback from the healthcare community) or indirectly by integrating them with regulatory or reimbursement mechanisms that stimulate actors to remain a part of the platform and/or to continuously use the offered e-health services over time. In either case, effective use of an intermediary platform will necessitate the active involvement of authorities, forward-looking decision makers, or opinion leaders.

An important precondition for entrepreneurs, SMEs, and firms joining the intermediary platform is the identification of who will have the financial responsibility for the services. It is today not clear if the services will be offered as public goods (and consequently be free for all individuals that are willing to use them) or if the services will be subsidized or provided in an almost "real" market structure. What is needed is a clear categorization of the types of e-health services which will remain the responsibility of public organizations and which parts of the market will be opened for entrepreneurs and innovative firms.

CUSTOMER ORIENTATION AS KEY TO SUCCESS

It is a strong first signal when a government commits to investing in e-health services and corresponding intermediary platforms. However, as in any service industry, only a limited level of success is possible when the customer is not onboard. The importance of customer orientation as an antecedent for e-health implementation success is supported by evidence from many research studies (Au et al., 2002; Nevo and Wade, 2007). However, value expectations of customers evolve over time (Sprenger et al., 2016). Therefore even if the impact of a two-sided market does not yield to instant revenues at the short term, an e-health service nevertheless might be successful in the long-term. For this to happen, the e-health service needs to equally deliver value to service consumers (e.g., care providers, citizens) and service providers (e.g., governmental authorities, private organizations). These can often be connected to improved welfare, patient security, improved public health, and increased commitments of consumers.

A central task is to actively investigate what service consumers actually want. The lack of knowledge on citizens' willingness-to-pay for which kind of e-health services, their individual preferences and attitudes toward value cocreation (McColl-Kennedy et al., 2012), or knowledge about the inter-temporal use of e-health services, is a major hurdle to entering the market. In this sense, a considerable effort needs to be done in the future to better understand and integrate customer orientation into the design and deployment of e-health services and intermediary platforms.

THE ROLE OF POLICY INNOVATION AND RENEWAL

Worldwide, policies appear to be in a state of constant reform. The immediate goal of many recent policy reforms has been to slow the growth of health spending by introducing changes and innovations, such as developing platforms supporting the interaction between patients, doctors, and other service providers in order to move from physical to digital infrastructures. By increasing the use of digital services and stimulating the growth of the market for e-health, a strong urge for reducing public expenditure is envisaged. However, an important policy question is the extent to which approaches should be actively encouraged by health reform legislations, and how such approaches should be organized and held accountable for. In this book we mention the call for innovative systems, which need to be established in order to satisfy consumers, providers, and policy makers alike, such that an effective renewal of the health system can take place. But an important area of contention remains whether the two-sided e-health market, which appears to be evolving in many places, is the appropriate way to achieve the envisaged objectives, or if alternative social or technological innovations (apart from e-health services) are required. Furthermore, many questions need to be answered, such as whether the market will be reduced to applications and services for long-term patients or patients with chronic diseases only, or if it will be an intermediary inclusive platform that will be sustained and owned by an organization that has the power to influence budget and allocation of resources and so on. All these questions need to be answered and a paradigm shift of strategies, practices, and guidelines as well pragmatic approaches to measure outcomes and the value and contribution to effectiveness and efficiency of a two-sided e-health market need to be introduced in many health systems all over the world. Additionally, clarity in terms of market regulation needs to be achieved today, as it often impedes SME, entrepreneurs, and other companies in getting active in the two-sided market. When the regulations of a market are changed, power is transferred from one actor to another (e.g., from a governmental authority to a private company or from the service provider to the service consumer).

To renew or remove regulations in order to establish an efficient two-sided market is therefore an issue that in some cases needs an adequate set of regulations and institutions that control if they are abided by. The formulation of these regulations is essential for the further development of a two-sided market and issues related to pricing, productivity, investment, and cost-effectiveness.

THE ROLE OF ENTREPRENEURIAL COLLABORATION

Renewal processes or the seeking of new future intermediate platforms are fostered by pragmatic entrepreneurial collaboration with visionary individual actors, company representatives, or academics, who like to challenge the existing capability limitations of today's health systems and institutions. In a two-sided e-health market, it is necessary to reducing transaction costs and overhead costs between the distinct actors by facilitating strong collaborations. The extent to which an intermediary platform influences the pricing structure, or the relative prices charged to distinct customer groups for accessing the platform or for the offered services, is of paramount importance by nature.

Prior research suggests that two-sided markets should charge prices which either just cover site-specific costs (and therefore do not contribute to overall profitability given that these platforms often have significant fixed costs), or which provide for the maintenance of services at below marginal costs. In a two-sided e-health market, the economic effects are related to the market power that an owner of intermediary platforms has, the installed base (or costumers of the intermediary platform), and the asymmetry of information that exists between the various stakeholders active in the market.

A two-sided e-health market faces competition from multinational firms with expedient pricing and profit structures for large proportions of the market, and single entrepreneurs and SMEs that render services for particular customer groups or market segments or service providers specialized in the support of intermediation (e.g., by providing and analyzing market information, running and maintaining platform services, and so on). An intermediary platform in a two-sided market may bundle features that provide value to both sides of the market. It is therefore necessary to identify issues that maximize social welfare, such as anticompetitive practices, exclusive contracting, or other protectionist approaches. Furthermore, the intermediary platform has to be open for all sorts of customer environments, whether it be single homes, multihomes, or institutional environments. In order to do that it is necessary that the intermediaries enable and stimulate entrepreneurial collaboration between the different actors on the platform. They must take over the role as "switch tenders" in that they operate in a flexible yet—to some extent—projectable way. This ability, in turn, implies that today's actors in the e-health market must radically change their organizational culture and behavior patterns from mainly administration-oriented to more entrepreneurship-oriented institutions. However, this transition cannot be taken lightly. It is a demanding challenge to launch innovative public procurement and private–public partnerships and to establish bottom-up processes and active dialogs with the business community. In this regard, the organization of (online and most importantly also offline) social networks composed of medical technology, pharmaceutical companies, and healthcare providers, such as governmental agencies, private and public hospitals, managed care centers, general practitioners, as well as other relevant professional groups in the e-health arena, is an issue of relevance for the future yet to come.

CONCLUDING REMARKS

We hope that this book has been of value to you. To our knowledge, it is the first book that deals with two-sided markets in e-health. With this book we employed a particular lens to explain success and failure in the e-health arena. Our center of attention has therefore been on the necessity of creating an ecosystem and corresponding strategies that sustain a type of market in which services are delivered online, across distinct organizational or even national borders. We are aware, however, that sustainable and innovative market structures only work when e-health services themselves are forward-looking and resourceful. In this sense, ecosystem development and service design need to go hand in hand. Accordingly, we expect that researchers and business professionals alike will enlighten us with additional, fresh insights on how to integrally design, maintain, and promote ecosystems and corresponding services. At the same time, we hope that we also could provide you with some novel ideas that enable you to move your research to deployment and hopefully to the market.

REFERENCES

Aanestad, M., Jensen, T.B., 2011. Building nation-wide information infrastructures in healthcare through modular implementation strategies. J. Strateg. Inf. Syst. 20 (2), 161–176.

Anderson, J.G., 2007. Social, ethical and legal barriers to e-health. Int. J. Med. Inform. 76 (5–6), 480–483.

Au, N., Ngai, E., Cheng, E., 2002. A critical review of end-user information system satisfaction research and a new research framework. Omega 30 (6), 451–478.

Connell, N.A.D., Young, T.P., 2007. Evaluating healthcare information systems through an "enterprise" perspective. Inf. Manage. 44 (4), 433–440.

McColl-Kennedy, J.R., Vargo, S.L., Dagger, T.S., Sweeney, J.C., van Kasteren, Y., 2012. Health care customer value cocreation practice styles. J. Serv. Res. 15 (4), 370–389.

Nevo, D., Wade, M.R., 2007. How to avoid disappointment by design. Commun. ACM 50 (4), 43–48.

Scott, T., Rundall, T., Vogt, T., Hsu, J., 2006. Kaiser Permanente's experience of implementing an electronic medical record: a qualitative study. Br. Med. J. 331 (3), 1313–1316.

Sprenger, M., Mettler, T., Winter, R., 2016. A viability theory for digital businesses: exploring the evolutionary changes of revenue mechanisms to support managerial decisions. Inf. Syst. Front., 1–24.

Vargo, S.L., Lusch, R.F., 2008. Service-dominant logic: continuing the evolution. J. Acad. Mark. Sci. 36 (1), 1–10.

AUTHOR INDEX

Note: Page numbers followed by "*f*" and "*t*" refer to figures and tables, respectively.

SUBJECT INDEX

Note: Page numbers followed by "*f*" and "*t*" refer to figures and tables, respectively.

Printed in the United States
By Bookmasters